Global warming policy in Japan and Britain

Manchester University Press

Issues in Environmental Politics

series editors Mikael Skou Andersen and Duncan Liefferink

At the start of the twenty-first century, the environment has come to stay as a central concern of global politics. This series takes key problems for environmental policy and examines the politics behind their cause and possible resolution. Accessible and eloquent, the books make available for a non-specialist readership some of the best research and most provocative thinking on humanity's relationship with the planet.

already published in the series

Science and politics in international environmental regimes
Steinar Andresen, Tora Skodvin, Arild Underdal and Jørgen Wettestad

Animals, politics and morality (2nd edn) *Robert Garner*

Implementing international environmental agreements in Russia
Geir Hønneland and Anne-Kristin Jørgensen

Implementing EU environmental policy *Christoph Knill and Andrea Lenschow (eds)*

Sweden and ecological governance: straddling the fence
Lennart J. Lundqvist

North Sea cooperation: linking international and domestic pollution control *Jon Birger Skjærseth*

Climate change and the oil industry: common problem, varying strategies *Jon Birger Skjærseth and Tora Skodvin*

Environmental policy-making in Britain, Germany and the European Union *Rüdiger K. W. Wurzel*

Global warming policy in Japan and Britain

Interactions between institutions and issue characteristics

Shizuka Oshitani

Manchester University Press

Manchester and New York

distributed exclusively in the USA by Palgrave

Published by Manchester University Press
Oxford Road, Manchester M13 9NR, UK
and Room 400, 175 Fifth Avenue, New York, NY 10010, USA
www.manchesteruniversitypress.co.uk

Distributed exclusively in the USA by
Palgrave, 175 Fifth Avenue, New York,
NY 10010, USA

Distributed exclusively in Canada by
UBC Press, University of British Columbia, 2029 West Mall,
Vancouver, BC, Canada V6T 1Z2

British Library Cataloguing-in-Publication Data
A catalogue record for this book is available from the British Library

Library of Congress Cataloging-in-Publication Data applied for

ISBN   0 7190 6938 6   hardback
EAN   978 0 7190 6938 3

First published 2006

15 14 13 12 11 10 09 08 07 06   10 9 8 7 6 5 4 3 2 1

Typeset in Sabon
by R. J. Footring Ltd, Derby
Printed in Great Britain
by Bell & Bain, Glasgow

Contents

Figures

Tables

Boxes

Acknowledgements

This study is based on my PhD thesis at the University of Essex, where I started the work in the mid-1990s. I am indebted to many people for all kinds of help in finally publishing it. Special thanks go to Albert Weale. He initially encouraged me to undertake an Anglo-Japanese comparison of environmental policy, an area which had previously been little researched. He provided me with various sources of information, valuable advice and encouragement. Without his thorough and patient support, this work would not have come to fruition. I would also like to express my sincere thanks to Kazuo Oshima, who supported me in various ways to continue the work further. Ian Neary, Shinichiro Murakami, Hugh Ward, Vicky Randall, Frances Millard and Stephen Wilks also provided me with comments and information, sometimes very specific. Duncan Liefferink read through the draft of this book and gave me several important suggestions for improvements.

I am also grateful to the various people in government, environmental groups and industrial associations, academics and politicians who kindly agreed to be interviewed by me and generously provided me with materials. Their information about political processes and their 'world views' greatly aided my understanding of climate policy and politics. Discussions with them were also a valuable source of new ideas.

I wish to thank Manchester University Press for reviews and comments, particularly Anthony Mason, a man of great patience.

Finally, I want to thank my family. Glenn was patient about my lingering work. Constant encouragement and support, especially from my parents, on various aspects enabled me to finalise this work. All the good ideas belong to them and others too numerous to mention; the mistakes are all mine.

Abbreviations

ACE (Britain)	Association for the Conservation of Energy
ACE (Japan)	Advisory Committee for Energy
ANRE	Agency for Natural Resources and Energy (Japan)
AOSIS	Alliance of Small Island States
CASA	Citizens' Alliance for Saving the Atmosphere and the Earth
CBI	Confederation of British Industry
CCL	climate change levy
CCLA	climate change levy agreement
CEGB	Central Electricity Generating Board
CFC	chlorofluorocarbon
CH_4	methane
CHP	combined heat and power
CNG	compressed natural gas
CO_2	carbon dioxide
COP	Conference of Parties (of the Framework Convention on Climate Change)
DEFRA	Department for Environment, Food and Rural Affairs
DEn	Department of Energy
DETR	Department of the Environment, Transport and Regions
DOE	Department of the Environment
DOT	Department of Transport
DTI	Department of Trade and Industry
EA	Environment Agency, Government of Japan
EEF	Engineering Employers' Federation
EEO	Energy Efficiency Office
EP	Energy Paper

EST	Energy Saving Trust
ETG	Emissions Trading Group
ETSU	Energy Technology Support Unit
EU	European Union
FCCC	(United Nations) Framework Convention on Climate Change
FOE	Friends of the Earth
G7 (8)	Group of Seven (Eight)
G77	Group of 77 (developing countries)
GAT	green automobile taxation
GDP	gross domestic product
GEN	Green Energy law Network
GHG(s)	greenhouse gas(es)
HEES	Home Energy Efficiency Scheme
HFC	hydrofluorocarbon
HMSO	Her Majesty's Stationery Office
IEA	International Energy Agency
INC	Intergovernmental Negotiating Committee (for a Framework Convention on Climate Change)
IPCC	Intergovernmental Panel on Climate Change
IPPR	Institute for Public Policy Research
JUSCANZ	Japan, the USA, Canada, Australia and New Zealand
Keidanren	Japan Federation of Economic Organisations
LDP	Liberal Democratic Party (Japan)
LEV	low-emission vehicle
LNG	liquefied natural gas
LPG	liquefied petroleum gas
MITI	Ministry of International Trade and Industry
MOC	Ministry of Construction
MOF	Ministry of Finance
MOFA	Ministry of Foreign Affairs
MOT	Ministry of Transport
MP(s)	Member(s) of Parliament
MW	megawatt
NEDO	New Energy and Industrial Technology Development Organisation
NEF	New Energy Foundation
NETA	New Electricity Trading Arrangements
NFFO	Non-Fossil Fuel Obligation

NGO	non-governmental organisation
NIES	National Institute for Environmental Studies
N_2O	nitrous oxide
NO_x	nitrogen oxides
ODA	Official Development Assistance (Japan)
OECD	Organisation for Economic Cooperation and Development
Ofgas	Office of Gas Supply
Ofgem	Office of Gas and Electricity Markets
P&R	pledge and review
PURPA	Public Utility Regulatory Policy Act (USA)
QMV	qualified majority voting
R&D	research and development
RCEP	Royal Commission on Environmental Pollution
RD&D	research, development and demonstration
RECs	regional electricity companies
RPS	renewable portfolio standards
RSPB	Royal Society for the Protection of Birds
SACTRA	Standing Advisory Committee on Trunk Road Assessment
SEA	Single European Act
TDM	transportation demand management
TSO	The Stationery Office
UN	United Nations
VAT	value added tax
WG	Working Group (of the IPCC) (there are three of these: WG-I, WG-II and WG-III)
WWF	World Wide Fund for Nature

1
Introduction

In 1987, in Bellagio, Italy, the possible problem of global warming was discussed for the first time at the *political* level since the notion of the human-induced greenhouse effect was suggested at the end of the nineteenth century. Then, after only two years of extensive diplomatic negotiations, more than 150 world leaders signed the Framework Convention on Climate Change (FCCC) at the United Nations Conference on Environment and Development held in Rio de Janeiro (the Earth Summit or Rio Summit) in 1992.

The significance of the Convention goes beyond the fact that the world agreed to cooperate to tackle the common threat to humankind. The most important greenhouse gas, carbon dioxide (CO_2), is generated by the burning of fossil fuels, which has been the foundation of industrialisation, civilisation and 'progress'. The Convention was an embodiment of the world realisation that the existing patterns of socio-economic policy and human activities were unsustainable. It represented the world's collective attempt to shift from a 'productivist' society – where social well-being and policy success are based on intensive economic growth (Lipietz, 1992) – to a 'sustainable' society, one which recognises the needs of future generations as well as those of present ones (World Commission on Environment and Development, 1987).

The adopted FCCC was a 'framework' Convention and did not offer specific policy to mitigate global warming. Moreover, many developed countries had already made specific commitments. Before the adoption of the more specific Kyoto Protocol at the end of 1997, under the FCCC policy developments were essentially left in the hands of each signatory. Some countries set more ambitious targets for the reduction of greenhouse gases than others, other countries introduced economic instruments to achieve their targets, and yet other countries formulated a carbon dioxide reduction programme that essentially consolidated policies that were already in place.

This book explores how different countries responded in the face of the common policy imperative of tackling global warming and why their responses were similar in some respects and different in others, taking Japan and Britain as examples. In-depth, systematic comparative environmental policy analysis of Japan and Britain is an area still little explored, despite the rapid growth in academic interest in comparative environmental politics and policy (Vogel, 1986; Boehmer-Christiansen and Skea, 1991; Hajer, 1995; Jänicke and Weidner, 1995; Wallace, 1995; O'Riordan and Jäger, 1996; Andersen and Leifferink, 1997; Lafferty and Meadowcroft, 2000a; Daugbjerg and Svendsen, 2001; Social Learning Group, 2001; Desai, 2002; Schreurs, 2002). One early work involving Britain and Japan is Enloe's (1975) comparison of pollution politics in the 1970s in four countries, the other two being the USA and the former USSR. Vogel also examined general environmental policy developments over time, and compared Japan and Britain against the USA in terms of the separation-of-power system in an analysis of institutional capacity for representing diffuse environmental interests (Vogel, 1990, 1993).

There has been little in the way of detailed Anglo-Japanese comparisons of environmental politics and policy, perhaps for two, not unrelated reasons: that, at first glance at least, differences between the two countries may appear subtle in terms of formal political institutions; and that Japan and Britain have somewhat similar histories and international reputations concerning their responses to environmental problems. For example, both countries have been relatively successful in tackling domestic air pollution. During the 1980s Japan was severely criticised for its relentless economic activities in other countries, which involved pollution exportation, forest destruction and over-fishing, while Britain was dubbed the 'dirty man of Europe'. Both countries were also viewed as reactive (as opposed to proactive) or negative on global environmental issues such as the depletion of the ozone layer. Then, towards the end of the 1980s, both started to adopt more positive stances (Schreurs *et al.*, 2001: 359–60). Regarding their formal political institutions, both countries are advanced liberal democracies, are centralised states in comparative terms, and use parliamentary systems. Neither country has fully fledged proportional representation. Britain essentially uses the single-member-district plurality system,[1] while Japan was and is, according to Lijphart (1999), a 'semi-proportional' system, both under the medium-sized multi-member-district system up to 1995 and since 1996 the system that combines the single-member-district plurality formula with proportional representation. Both countries have also been strong 'productivists'. For a long time after the Second World War, Britain was preoccupied with reversing its declining economic competitiveness, while Japan pursued rapid economic growth. Given these common features, one might consider it not surprising that the two countries took similar courses of action on the environment.

However, Japan and Britain are contrasting countries in other important respects. In his study on environmental representation in the USA, the USSR, Japan and Britain, Vogel gives an important rationale for Anglo-Japanese comparison. According to him, Britain is classified as 'a weak state with relatively little institutional capacity for shaping the structure and priorities of civil society', while Japan is 'a strong state with considerable institutional and political capacity to shape structure and priorities' (Vogel, 1993: 241). His point can be explicated in terms of political economy. In Japan, the government intervenes in markets and steers the economy (see Johnson, 1982). Planning is a key feature of state activity. Business is relatively well integrated and government and organised business collaborate with each other in making and implementing policies. In other words, Japan possesses important features of corporatism. Britain, on the other hand, has a strong liberal state tradition. Unlike in Japan, interventionism is alien in Britain. Society is considered to be composed of atomised, competing individuals pursuing their own interests, 'without reference to society and interference from the state' (Marquand, 1988). If Japan has important corporatist elements in government–market relations, Britain has elements of pluralism.

Japan and Britain are also contrasting in their prevailing norms of decision-making and the relations between the government and markets. In Britain, where the electoral system produces a two-party system, political power is centralised to the majority party. In Japan, until 1993, when Japan entered an era of coalition (or loosely allied) governments, the Liberal Democratic Party (LDP) enjoyed the most stable one-party dominance of any advanced liberal democracy in the postwar era. Notwithstanding, the LDP tended to follow 'the norm of cross-party consensus building' (Pempel, 1992). Thus, even before 1993, Japan's decision-making style was consensual rather than 'majoritarian' (the latter characterises the British mode of legislative decision-making).

These two institutional features which contrast Japan and Britain are not unrelated to each other. According to Crepaz and Lijphart (1995), both corporatist interest-group politics and consensus-based legislative policy-making are structured by norms of cooperation and collective decision-making by means of accommodation; in contrast, pluralism and majoritarian legislative democracy are characterised by competition and a 'winner takes all' decision-making style. Not only do corporatism and consensus democracy, and pluralism and majoritarian democracy have a conceptual affinity with each other, but their links are also empirically supported. Japan and Britain can be, thus, posited as having the contrasting institutional features of combined corporatism–consensus policy-making and combined pluralist–majoritarian policy-making, respectively.[2] Although broadly similar, Japan and Britain have important institutional differences therefore in style of policy-making (Table 1.1).

The following questions then arise in the context of the present work. Does this contrast make any difference to Japanese and British responses to the problem of global warming? And if so, how? What is beneath these questions is a theoretical enquiry into the role of institutions in politics and policy. Do institutions 'matter'? If so, how and when do they 'matter'? These questions have been around for a while in political science. There are reasons, however, why the problem of global warming may throw new light on them. First, the problem is unique in its cause–effect relations. Today's costly action will bring its beneficial effects decades or even centuries later. This will pose critical constraints on policy-makers across countries, as they operate and are oriented towards a much shorter time frame. Second, the problem is truly global and therefore any action taken (which will be costly) could be undermined by 'free-riders', who could thereby deny any benefit accruing to the countries that take action. Third, there remains uncertainty over the scale and timing of global warming, as well as its specific

Table 1.1 *Japan and Britain compared and contrasted (selected features)*

	Japan	Britain
Productivism	Primacy of economic growth	Primacy of recovering competitiveness
Federal or unitary	Unitary, centralised	Non-federal, relatively centralised
Executive–legislative relations	Parliamentary system	Parliamentary system
Proportional representation electoral system?	No, semi-proportional representation	No, plurality formula
Legislative decision-making styles	Consensus	Majoritarian
Government–market (society) relations	Strategic interventionism with corporatist characteristics	Non-interventionism and liberal state tradition with pluralist characteristics

effects. The conventional cost–benefit analysis in policy-making is therefore difficult to employ. All in all, the problem of global warming poses an unconventional challenge to policy-makers.

In analysing global warming policy in Japan and Britain and exploring the theoretical enquiry into the role of institutions, by drawing on Freeman's (1985) contrasts between national styles and policy sectors, I will put forth two theoretical perspectives. One is the institutional approach, which emphasises the importance of national institutions in shaping politics and policy. The other is the issue-based approach, which emphasises the constraints inherent in an issue that fall on rational actors regardless of the institutions in which they are operating. From this perspective, it is this configuration of inherent constraints that determines political interactions, and hence policy outcomes. The implication of the two approaches is clear. While the institutional approach predicts policy differences between different countries, the issue-based approach indicates the possibility of similarities in both policy and outcome. Freeman, however, suggests that the two approaches should not be treated as mutually exclusive; rather, they need to be integrated 'in an appropriately complementary manner' (Freeman, 1985: 469). Rather than examining which approach is right or better, therefore, I aim to combine the issue-based approach with the institutional approach in analysing Japanese and British policies on global warming and to explore the relationships between the two approaches, and hence the role of institutions in explaining politics and policy.

The comparative analysis of policies on global warming in Japan and Britain can open up another question. If institutions matter, the next question is, as indicated above, 'How?' Having characterised Japan, institutionally, as adopting consensus corporatism and Britain majoritarian pluralism, we can explore the effect of levels of corporatism on environmental policy and performance. Corporatism's effects on the environment have recently attracted growing interest among environmental and comparative policy analysts. It is argued that corporatism's interventionism and practice of policy concertation[3] are institutional assets for successful environmental policy-making (OECD, 1993; Jänicke, 1995; Dryzek, 1997). On the other hand, some argue that corporatist consensus-seeking through elitist bargaining results in short-termism, making it difficult for important long-term environmental interests to come into the centre of policy-making (Opschoor and Straaten, 1993; Hukkinen, 1995a,b). Does a corporatist policy-making process make it easier or harder to tackle environmental challenges? A comparative analysis of Japanese and British policies on global warming will make a contribution to this debate.

'Policy' is a broad and multidimensional concept. I will look specifically at four aspects of it, as to break policy down in this way will help us to understand whether policies in the two countries are similar or different and to make analysis systematic. The four aspects to be looked at are:

- the speed of policy change;
- the content of policy, including the choice of policy instruments;
- the degree of integration of global warming concerns into the policy areas of energy and transport;
- policy stringency.

The speed of policy change gives insight into how responsive a country is to a particular policy imperative. Policy contents, including the choice of policy instruments, are the means by which goals are achieved. This is an important aspect of policy which is closely related to core beliefs embedded in policy (Hall, 1993). Policy integration highlights the processes of changes (or not) over time. This is an important reference point for assessing progress on global warming policy. Here, 'qualitative' progress is of more interest than 'quantitative' progress (which could be measured in terms of reductions in emissions, which are necessarily affected by a large number of external, socio-economic factors and crucially by emission levels at the point when the issue arose on the agenda). Last, policy stringency indicates policy-makers' degree of ambition as well as their capability to pursue particular goals. Together, these four aspects of policy will provide a good vantage point for understanding a country's policy on global warming.

The main argument of this book is that although the institutional approach is useful in explaining certain aspects of global warming policy in Japan and Britain, the issue-based approach gives crucial insights into their policies and the fundamental constraints on policy-makers developing global warming policy. It is argued that institutionalists need to pay due attention to the configuration of constraints that an issue poses to rational agents who act in specific institutional settings.

The organisation of the book

Chapter 2 examines both the issue at hand and the international politics which evolved in response to it. What is the problem of global warming? What are its implications for domestic policy? What is the FCCC? The chapter provides the basic information necessary for the ensuing explanation of Japanese and British global warming policy.

Chapter 3 then delineates the theoretical framework for the analysis. As mentioned above, I use two approaches to comparative politics and policy: the institutional approach and the issue-based approach. The institutions that I focus on are those related to consensus corporatism and majoritarian pluralism. The implications that the institutional and the issue-based approaches have for global warming policy in Japan and Britain are discussed. In addition to the two approaches, international influences and individual policy catalysts are also discussed, but in this book these two variables are treated essentially as exogenous stimuli to domestic policy processes.

Chapter 4 discusses styles of environmental policy-making in Japan and Britain. In so doing, I will elaborate the argument that Japan has an consensus corporatist institutional character while Britain has a majoritarian pluralist character. The argument about institutions and policy-making styles is followed by an examination of the economic contexts of global warming politics in Japan and Britain, and descriptions of the main actors involved in the policy process.

Chapters 5–8 describe the politics of global warming in Japan and Britain. Chapters 5 and 7 look at global warming policy developments in the two countries, while Chapters 6 and 8 look at policy integration. The latter examine how and to what extent the climate message was integrated into energy and transport policy. Energy and transport policy areas are chosen because of their importance to global warming. They are also the areas which symbolise the present mode of 'progress' (von Weizsäcker, 1994). As such, policy integration in these sectors provides a touchstone of the transition from 'productivism' to 'sustainability'. There are inevitable overlaps between Chapters 5 and 6, and between Chapters 7 and 8. Apart from a practical reason pertaining to the length of each chapter, the reason for dividing the material for each country into two chapters is that the analysis of policy integration, which treats the issue of global warming as a *cause*, entails different perspectives from the analysis of global warming policy, seen as an *effect*. Dividing the material for each country allows each chapter to serve its own purpose and helps make this thesis clearer and more focused, thus making analysis easier.

Chapter 9 brings together the threads from the empirical chapters to compare, contrast and analyse Japanese and British policies using the theoretical framework provided in Chapter 3. On the basis of the findings I will further discuss the questions raised in this Introduction, namely concerning the interaction between the institutional and issue-based approaches, and the relations between the levels of corporatism and the institutional capacity to tackle the challenges of sustainable development. In Chapter 10, the Epilogue, I will briefly look at major policy developments in Japan and Britain since the adoption of the Kyoto Protocol and consider their implications for the main findings of this book.

This study looks mainly at policy developments between 1988 and 1997, that is, since the emergence of the issue of global warming on the international and national agenda, up to the point at which more specific international influences on domestic policy came with the adoption of the Kyoto Protocol in December 1997 at the Third Conference of Parties to the Framework Convention on Climate Change. The period this book covers is, as it were, the first stage of climate politics, which is now followed by the second stage, lasting up to 2008–12 (the first commitment period under the Kyoto Protocol). The pre-Kyoto period is a critical one, when politics and policy developed which, in one way or another, conditioned future paths. Analysis of the politics and policy characterising the first stage would certainly be of use for further research on climate politics beyond the second stage. In Britain, this first stage of climate politics largely falls within a period of uninterrupted Conservative government. The Labour Party came to office in May 1997, but by this time the British stance at the Rio Summit was decided under the framework of the European Union (EU) and the major domestic policy developments under Labour took place largely after the adoption of the Kyoto Protocol. The change of government necessarily raises the question of the role of ideology in climate policy. I will examine this question in the Epilogue, which looks at major policy developments in Japan and Britain after the adoption of the Kyoto Protocol.

At the time of writing (January 2005), international climate politics is about to take the long-awaited step forward, with the Kyoto Protocol due to enter into effect in February 2005. This will certainly give strong support to the advocates of robust climate policy in Japan and Britain; however, it is still hard to see what developments (or otherwise) the Kyoto Protocol's entering into effect will bring about.

Although global warming policy should include adaptation to the possible consequences of global warming, abatement of all anthropogenic greenhouse gases and the enhancement of 'sinks', that is, reservoirs of greenhouse gas emissions such as forests and soils, I will focus on policies regarding the abatement of emissions of carbon dioxide. As explained in the next chapter, this is because the limitation of carbon dioxide emissions is widely considered central to policy on global warming. Also, the carbon dioxide problem alone is rich enough in policy and politics to be the focus of a whole book. The terms 'carbon dioxide policy', 'global warming policy' and 'climate (change) policy' are therefore used interchangeably in this study.

It is worth noting that in both Japan and Britain, the government administration was reorganised or reformed after 1997. In Japan in 2001 the number of government ministries and agencies was almost halved, mainly through radical mergers, and new names and tasks were given to them. Within this reorganisation, the Environment Agency (EA) was the only body which was explicitly expanded, to become the Ministry of the Environment. In Britain under the incoming Labour government the Department of the Environment underwent two reorganisations, with the creation of the Department of the Environment, Transport and the Regions in 1997 and the Department for Environment, Food and Rural Affairs in 2001. Given the time frame of this book, the ministerial or departmental names before these reorganisations are used.

Notes

1 This system is used for parliamentary and local authority elections. Elections to the devolved assemblies of Wales, Scotland and Northern Ireland use systems of proportional representation, as do elections to the European Parliament.

2 I am aware that the depiction of Japan as corporatist is open to contention, but, as argued in Chapter 3, in terms of analysing environmental policy-making, in particular, Japan has the important features of corporatism as opposed to those of pluralism.

3 In this book policy 'concertation' is broadly defined as cooperative policy-making and concerted action (orchestrated by government) towards the policy goals agreed upon between government and representatives of affected groups.

2
Science and the international politics of global warming

This chapter gives a brief overview of: the nature of the problem of global warming; the international political responses to the scientific developments; and the international regime on global warming, the United Nations FCCC.[1] Grasping the nature of the problem is an important first step towards understanding national policy and policy developments, since any problem poses a certain set of constraints on the choices of rational and strategic policy-makers. Nor can we afford to overlook international political developments when we examine transboundary problems, which inevitably erode the political significance of national territory.

The IPCC and the science of global warming

The state-of-the-art scientific knowledge of global warming has been given by the Intergovernmental Panel on Climate Change (IPCC), which was established in 1988. The IPCC is composed of three Working Groups (WGs), to which leading scientists and technical experts are nominated by their governments and by international organisations, including non-governmental organisations (NGOs). WG-I assesses the science of global warming. Both WG-II and WG-III have been reorganised but basically look at the effects of global warming, response strategies and their implications. The IPCC has produced three assessment reports (Houghton *et al.*, 1990; Tegart *et al.*, 1990; IPCC, 1991, 1996, 2001a); the fourth assessment report is expected in 2007. The IPCC has also produced special reports and technical papers as requested by parties to the FCCC and/or with its own initiative. What is special about these reports is their authority. The reports represent a broad consensus within the expert community.[2]

Scientific work by the IPCC has greatly contributed to elucidating the mechanism of global warming, but the basic picture has not changed much since 1990.

What is the greenhouse effect?

The greenhouse effect itself is a natural phenomenon. Its mechanism is basically as follows (see Figure 2.1 for a simplified illustration). Incoming solar radiation at short wavelengths passes through the clear atmosphere, and most of it is absorbed by the earth's surface, although some is reflected by the earth and the atmosphere and radiated back into space at longer wavelengths. Although the atmosphere absorbs almost no short-wave energy from incoming solar radiation, the trace quantities of greenhouse gases (GHGs) in the atmosphere retain the reflected long-wave energy, thus warming up the earth. The main natural GHGs are water vapour, carbon dioxide (CO_2), methane (CH_4), nitrous oxide (N_2O) and tropospheric ozone (O_3). Together they account for only 1 per cent of

Figure 2.1 *The mechanism of the greenhouse effect.*

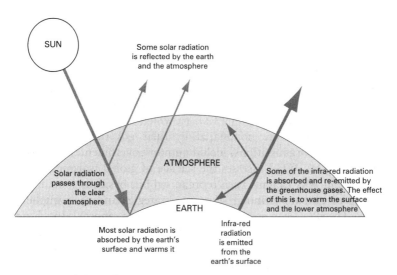

Source: Houghton *et al.* (1992: 7).

Figure 2.2 *Contribution of global greenhouse gases to the enhanced greenhouse effect. PFCs = CF₄ and C₂F₆. HFCs = HFC-23, HFC-134a, HFC-152a.*

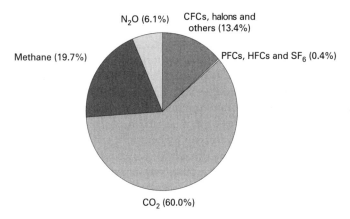

Source: IPCC (2001a: 356–8).

the atmosphere, but they have kept the earth warm enough to be habitable. The problem of global warming is the anthropogenic, 'enhanced' greenhouse effect – over and above the natural phenomenon. Since the Industrial Revolution, human activities have added these and new GHGs at an unprecedented rate and volume. The resultant changes in the composition of the atmosphere raised fears of global warming, or climate change generally.[3]

The relative contribution of GHGs to the 'enhanced' greenhouse effect is shown in Figure 2.2. Carbon dioxide is the most important, responsible for 60 per cent of the enhanced greenhouse effect, followed by methane (20 per cent) and nitrous oxide (6 per cent). It should be noted that these gases are all long-lived, so that their warming effect is both global and enduring.[4]

Table 2.1 presents the characteristics of the main GHGs (covered by the Kyoto Protocol, as explained below). As mentioned above, one of the important properties of the main GHGs is their longevity. An important policy implication is that there is a considerable time lag between their emission or a reduction of emissions and the effects. Thus, even if carbon dioxide emissions are stabilised immediately, it would take 100–300 years for concentrations to stabilise, approaching a level double the pre-industrial concentration. Moreover, inertia in climate systems means that even if atmospheric concentrations of GHGs are stabilised, the global temperature would continue to rise for a few centuries, and the sea level would continue to rise for many centuries to millennia (IPCC, 2001b: 17). Another point to note is that the principal cause of GHG emissions is the burning of fossil fuels, which has been the foundation of industrialisation and civilisation. Among fossil fuels, coal is the most carbon intensive, followed by oil and then gas. No feasible technology is currently available to remove carbon dioxide emissions from the burning of fossil fuels. Addressing global warming entails a radical move away from the existing economic system and civilisation based on massive use of fossil fuels.

Since its first assessment report, published in 1990, the IPCC has reported rapid increases in concentrations of the main GHGs and the picture of a warming world (see Figures 2.3 and 2.4 for changes since the Industrial Revolution and Figure 2.5 for an illustration of the change over the past 160,000 years). According to the IPCC, during the twentieth century, global mean temperature rose by about 0.6°C and global average sea level by 10–20 cm (IPCC, 2001a: 2–4). The question is, however, whether and to what extent human activities are relevant to this warming. On this point, the IPCC's scientists have become more confident that 'enhanced' global warming is taking place. Its third assessment report states that 'An increasing body of observations gives a collective picture of a warming world and other changes in the climate system', with 'new and stronger evidence that most of the warming observed over the last 50 years is attributable to human activities' (IPCC, 2001a: 2, 10).

The IPCC's scientists have also predicted the effects of continued emissions of GHGs from human activities on the atmosphere and climate. According to the third assessment report, if no action is taken, the global average temperature will rise by about 1.4–5.8°C between 1990 and 2100 – a rate of change very likely to be without precedent during at least the last 10,000 years. This would raise the global average sea level by about 0.09–0.88 metres over the same period (IPCC, 2001a: 13–16), which would pose serious risks to small islands and low-lying coastal areas. A 1 metre rise would displace 70 million people in Bangladesh, for example, and submerge 80 per cent of the Marshall Islands. Salt

Table 2.1 Summary of key anthropogenic greenhouse gases (except for those under the Montreal Protocol and its amendments[a])

	CO_2	CH_4	N_2O	HFCs (HFC-23)	PFCs (CF_4)	SF_6
Pre-industrial concentration	280 ppm	700 ppb	270 ppb	0	40 ppt	0
Current concentration	365 ppm	1,745 ppb	314 ppb	14 ppt	80 ppt	4.2 ppt
Rate of concentration change[b] (per year)	1.5 ppm	7.0 ppb	0.8 ppb	0.55 ppt	1 ppt	0.24 ppt
Atmospheric lifetime	5–200 years[c]	12 years	114 years[d]	260 years	>50,000 years	3,200 years
Principal anthropogenic sources	Fossil fuels (80%); deforestation and land-use change (10–30%)	Fossil fuels (20%); livestock (30%); wetland or paddy rice (20–25%)	Agricultural soil (61.4%); industrial sources (22.8%); biomass burning (8.7%)	Ozone-safe replacements for CFCs	Ozone-safe replacements for CFCs; a by-product of aluminium smelting	An electric insulator, heat conductor and freezing agent
Reductions required for stabilisation of the atmospheric concentration[e]	Immediate reduction of 50–70% and further reduction thereafter	Immediate reduction of about 8%	Immediate reduction of more than 50%	Not specified	Stop emissions	Stop emissions

HFCs, hydrofluorocarbons; PFCs, perfluorocarbons; SF_6, sulphur hexafluoride.

ppm = parts per million; ppb = parts per billion; ppt = parts per trillion.

[a] The Montreal Protocol on Substances that Deplete the Ozone Layer, adopted in Montreal in September 1987 and as subsequently adjusted and amended. Ninety-six chemicals are presently controlled under the Montreal Protocol and its amendments, including CFCs and halons.

[b] The rate is calculated over the period 1990–99. The rate fluctuated between 0.9 ppm/year and 2.8 ppm/year for CO_2 and between 0 ppb/year and 13 ppb/year for CH_4 over the period.

[c] No single lifetime can be given for CO_2 because of the different rates of uptake by different removal processes.

[d] This lifetime has been defined as an 'adjustment time', to take into account the indirect effect of the gas on its own residence time.

[e] IPCC (1996: paras 4.6, 4.13–4.15). According to the IPCC's third assessment report (2001a: 38, 244), stabilisation of atmospheric CO_2 concentrations at 450, 650 or 1,000 ppm would require emissions to drop below 1990 levels, within a few decades, about a century, or about two centuries, respectively, and to continue to decrease steadily thereafter. Eventually CO_2 emissions would need to decline to a very small fraction of current emissions. It is widely considered that stabilisation at 650 ppm and especially 1,000 ppm would significantly change the global climate.

Figure 2.3 *Indicators of the human influence on the atmosphere during the industrial era (global atmospheric concentrations of three well mixed greenhouse gases). Note that the ice core and fern data for several sites in Antarctica and Greenland (shown by different symbols) are supplemented with the data from direct atmospheric samples over the past few decades (shown by the line for carbon dioxide and incorporated in the curve representing the global average for methane). The estimated positive radiative forcing from these gases is indicated on the right-hand scale. Radiative forcing is 'an externally imposed perturbation in the radiative energy budget of the Earth's climate system' (IPCC, 1996: 353). A positive radiative forcing tends to warm the earth's surface.*

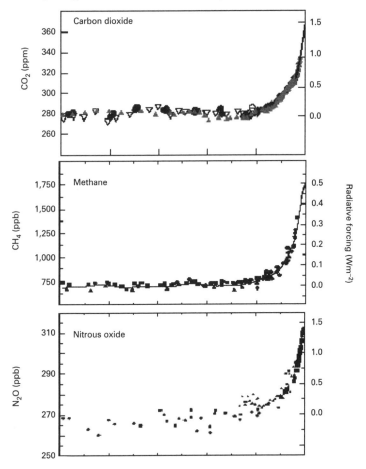

Source: IPCC (2001a: 6).

Figure 2.4 *Combined air and sea surface temperature anomalies (°C), 1861–2000, relative to 1961–90. Note that the bars on the annual number represent two standard errors.*

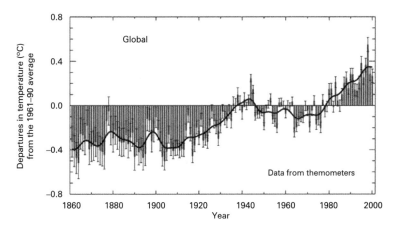

Source: IPCC (2001a: 26).

Figure 2.5 *Temperature and carbon dioxide concentrations over the last 160,000 years (from ice cores).*

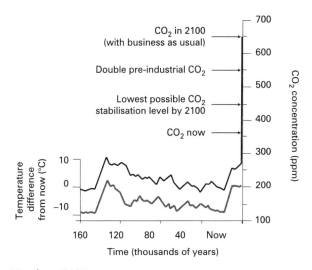

Source: Houghton (2002).

water could intrude into rivers and coastal areas, affecting fresh-
water supplies and fishing. Rapid rises in temperature and changes
in climatic pattern, including that of precipitation, would intensify
extreme climate events, and have various effects on human health
and infrastructure, biodiversity, food production, water resources
and economies. Both beneficial and adverse effects could result if
the warming is less than a few degrees Celsius, but adverse effects
would predominate for much of the world. In general, however,
negative effects are expected to fall most heavily on developing
countries and the poor within countries. If warming goes beyond
a few degrees Celsius, most regions would be at risk of predominantly negative effects, and there would also be the risk of abrupt
and large-scale 'surprises', such as the loss of a substantial fraction
of the total glacier mass beyond the next century (IPCC, 2001b).

There remain a number of scientific uncertainties, however. The
IPCC's scientists are not certain about the likelihood and details of
large-scale abrupt changes in climate (IPCC, 2001b: para. 9.16).
Moreover, the IPCC has yet to explain some past climatic phe-
nomena, which might cast doubt on its ability to predict climate
change accurately. Not least because of these uncertainties, some
scientists dispute the IPCC's view of the relevance of the human
influence via GHG concentrations to global warming that has
occurred since the Industrial Revolution. For sceptics, there are
too many uncertainties surrounding key issues to declare unequiv-
ocally that human activities are influencing climate.

Nevertheless, given the increasing evidence and broad consensus
on enhanced global warming among leading climate scientists and
given the risk of irreversible damage and cost, in the second assessment report, published in 1996, the IPCC concluded that 'the risk
of aggregate net damage due to climate change, consideration of
risk aversion and the precautionary approach, provide rationales
for action beyond "no regrets"' (para. 8.2).[5]

International political developments on climate change

The Villach–Bellagio workshops held by scientists in 1987 were a
turning point in the discourse on global warming. The discourse
entered the political arena. Participants in the workshops not only
confirmed global warming trends but also, for the first time, discussed and agreed detailed recommendations on policy to tackle

the possible problem of climate change. In October 1988 over 300 individuals (including climatologists, lawyers, public officials and private individuals) from forty-six countries attended the conference 'Changing Atmosphere: Implications for a Global Security' in Toronto (Rowlands, 1995: 74) and made a number of further recommendations, including the target of a 20 per cent reduction in carbon dioxide emissions from 1988 emission levels by 2005 (the 'Toronto target'), with an eventual aim of a 50 per cent reduction.[6] The establishment of the IPCC was also an important fruit of this conference (Paterson, 1992: 176).

A series of international conferences followed (see Table 2.2 for a summary of these) and by 1990 an international norm was emerging on the politically desirable level of carbon dioxide reduction. In November 1989 at the Noordwijk conference, in the Netherlands, representatives from seventy-two countries agreed on a declaration that demanded the stabilisation of carbon dioxide emissions by 2000 at the latest. Although the base year was not made clear owing to opposition from some countries (including Japan and Britain), the stabilisation of carbon dioxide emissions at 1990 levels by the end of the century was increasingly seen as a desirable first step. A new landscape of international political divisions also began to emerge. At one extreme were the oil-producing countries, which resisted any political developments. At the other extreme was the Alliance of Small Island States (AOSIS), composed of low-lying countries whose survival would be seriously threatened by rises in sea level and which were, hence, pushing the world to take action. Developed countries fell between these two extremes, but there emerged an increasing conflict between the positions adopted by a reluctant USA – the biggest contributor to GHG emissions in the world – and the EU, which was disposed to take positive action.

In May 1990, the IPCC published its first assessment report, which concluded that human-induced global warming was likely to be a reality. In November, the Second World Climate Conference formally decided to start negotiations on an international treaty on global warming, and in December the United Nations General Assembly established the Intergovernmental Negotiating Committee (INC) for a Framework Convention on Climate Change, with the aim of reaching an international agreement on a treaty before the Earth Summit in Rio in June 1992. In

Table 2.2 *The major international conferences in the year following the Toronto conference*

Place	Date	Conference	Main result(s)
Ottawa	February 1989	'International Meeting of Legal and Policy Experts on the Protection of the Atmosphere'	A statement on elements to be included in a framework convention on the protection of the atmosphere, including the creation of a World Atmosphere Trust Fund
The Hague	March 1989	A summit meeting on the protection of the atmosphere	A call for the formation of new global institutions to meet the challenges of global atmospheric change
Paris	June 1989	The G7 summit	Dubbed the 'green summit', it called for a 'framework or umbrella convention'
Noordwijk	November 1989	'Ministerial Conference on Atmospheric Pollution and Climate Change', attended by representatives from seventy-two countries	Agreement that industrialised countries should stabilise their CO_2 emissions at levels recommended by the IPCC

May 1992, after fifteen months of hard negotiations, the FCCC was adopted. At the Rio Summit in June, the leaders of most countries (the exclusions notably including the USA) signed the Convention. The Convention came into force on 21 March 1994, ninety days after the receipt of the fiftieth ratification. At the time of writing, 188 governments (including the EU) are parties to the FCCC (Climate Change Secretariat, 2002). (Details of the Convention are looked at in the next section.)

Currently the Conference of Parties (COP) of the FCCC meets once a year to monitor its implementation and keep the world effort going. Table 2.3 summarises the COPs to date.

The main objective of COP1 was to solve the problems which were left unresolved when the FCCC was adopted. The most important among them concerned the commitment beyond 2000. The indicative target for developed countries under the FCCC was only to 'return' GHG emissions in 2000 to the 1990 levels. Whether this was adequate to prevent climate change and what countries should do beyond 2000 were questions left to COP1.

Negotiations and discussions in advance of COP1 took place in an atmosphere of diminishing environmental zeal after the Earth Summit in 1992. Many developed countries were preoccupied with getting out of recession and were unlikely to meet their voluntary national GHG reduction targets towards 2000. Although before COP1 there had emerged a consensus on the inadequacy of existing commitments and a call for further commitments beyond 2000, countries were sharply divided on the extent to which they should go further. Some argued for a simple 'stabilisation' of GHG emissions, while AOSIS, building on a preliminary proposal for a protocol by Germany, sought a 20 per cent cut in carbon dioxide emissions by 2005. Some wanted legally binding targets and time-tables for reductions in GHG emissions, while others were strongly opposed to them. As a result, although the initial intention was to review and adopt new targets beyond 2000, as COP1 approached it became clear that its task was, at the very least, formally to recognise the inadequacy of the FCCC's commitment and, at best, to agree to a mandate to negotiate further commitments (Arts and Rüdig, 1995).

After heated negotiations, and helped by the formation of an innovative 'green group' – an alliance between the EU, the Group of 77 (G77)[7] and China – COP1 adopted the 'Berlin Mandate', which called on developed countries to create, for adoption at

Table 2.3 *The history of the Conference of Parties*

Session	Place and date	Main outcome
COP1	Berlin, March/April 1995	The 'Berlin Mandate' to launch a process towards appropriate action for the period beyond 2000, including the strengthening of the commitments of 'Annex I parties'[a] through the adoption of a 'protocol or another legal instrument'
COP2	Geneva, July 1996	The 'Geneva Ministerial Declaration', which endorsed the IPCC's conclusions and called for legally binding objectives and significant reductions in GHGs
COP3	Kyoto, December 1997	The Kyoto Protocol adopted, as promised in the Berlin Mandate
COP4	Buenos Aires, November 1998	The 'Buenos Aires Plan of Action' adopted, which addressed a number of issues to be elaborated under the FCCC and finalised the rulebook of the Protocol so that the agreement would come into effect at COP6
COP5	Bonn, October/November 1999	A timetable agreed for completing the details of the Kyoto Protocol by November 2000
COP6	The Hague, November 2000	Delegates failed to reach agreement on key issues under the Buenos Aires Plan of Action to make the Kyoto Protocol operational
Resumed COP6	Bonn, July 2001	The 'Bonn Agreement' adopted, registering consensus on the most politically controversial issues under the Buenos Aires Plan of Action
COP7	Marrakech, October/November 2001	The 'Marrakech Accords' adopted, a set of core decisions giving effect to the Bonn Agreement
COP8	New Delhi, October/November 2002	The 'Delhi Ministerial Declaration on Climate Change and Sustainable Development' adopted
COP9	Milan, December 2003	Progress on detailed rules for implementing the Kyoto Protocol
COP10	Buenos Aires, December 2004	An agreement on a seminar of governmental experts to be held in May 2005 in Bonn 'without prejudices to any future negotiations, commitments, process, framework or mandate under the Convention and the Kyoto Protocol'

[a] 'Annex I parties' are those listed in Annex I of the FCCC. They are those industrialised countries, including those with economies in transition, that have historically contributed the most to climate change.

COP3, a 'protocol or another legal instrument' that would set quantified emission limits with specified time frames, such as by 2005, 2010 and 2020. It is said that the adoption of the Berlin Mandate owed much to a special report from the IPCC (1994) whose conclusion indicated that the existing commitment was inadequate to prevent climate change (Takeuchi, 1998: 127).

COP2, held in Geneva in July 1996, was a 'halfway stage' between the Berlin Mandate and the adoption of a 'protocol or another legal instrument' at COP3 (Newell and Paterson, 1996). Progress was slow but COP2 produced an important political statement, the 'Geneva Ministerial Declaration'. The Declaration called for 'legally binding' objectives and 'significant' reductions, instead of 'limitation and reduction', in GHG emissions. Whether objectives were to be 'legally binding' had been the subject of controversy since the adequacy of the commitment had been under discussion. The incorporation of the 'legally binding objectives' in the Declaration was largely due to a sudden shift in position during COP2 by the USA. The USA had been, together with Japan, Canada, Australia and New Zealand (JUSCANZ standing for these five countries), strongly opposed to the legally binding objectives which were desired by the EU. The Geneva Ministerial Declaration was, however, not accepted by all the parties. Australia and New Zealand expressed reservations about such objectives, and fifteen oil-producing countries did not sign it.

The Geneva Ministerial Declaration endorsed the IPCC's call for an urgent strengthening of action and confirmed that the current levels of GHG emissions would result in 'dangerous interference with the climate', the avoidance of which is the FCCC's ultimate objective. Also, by noting the IPCC's finding that the stabilisation of carbon dioxide concentrations at 550 ppm (twice pre-industrial levels) by 2100 would eventually require global emissions to be less than half current levels, the Declaration implicitly set a desired global target. The IPCC has avoided endorsing any particular level of GHG concentration as 'safe', but the level of 550 ppm began to be considered by policy-makers practicable and necessary to prevent significant climate change (UNU–TERI Initiative on Climate Change, 1998). This 550 ppm objective was pressed by the EU during COP2. The EU had agreed at its Environment Council in June 1996 that levels lower than 550 ppm should guide limitation and reduction efforts.

Differences in positions, especially between the EU and the USA, narrowed little in the run-up to COP3. The central issue was the numerical objectives, but parties were divided not only on target figures but also on fundamental issues, such as which gases should be covered, when the target and baseline years should be, whether targets should be flat or differentiated among countries, and whether removal of carbon from sinks (forests, croplands and grazing lands, which absorb carbon in the atmosphere) should be counted. The world was also divided over the use of flexible mechanisms in achieving targets – banking and borrowing of emissions savings, emissions trading and joint implementation (see below). Against the Berlin Mandate, which promised no new obligations for developing countries, the role of developing countries in achieving the FCCC's ultimate objective was re-tabled. With too many key controversial issues unsettled, there was doubt about whether COP3 could successfully fulfil the Berlin Mandate.

Nevertheless, with significant compromises by all parties and skilful leadership of its chairman, Raúl Estrada Ouyela, COP3 adopted the Kyoto Protocol after an all-night session and a one-day extension of the schedule. The thrust of the Protocol is that developed countries shall reduce their overall emissions of six key GHGs by at least 5 per cent by the first commitment period (2008–12). Specific abatement targets for each country were politically decided. Emissions will be calculated by combining carbon dioxide, methane, nitrous oxide, hydrofluorocarbons, perfluorocarbons and sulphur hexafluoride in a 'basket', with reductions translated into carbon dioxide equivalents. As a means to allow governments to achieve national targets in an easier and more cost-effective manner, the 'Kyoto mechanisms' were introduced. Also called flexible or marked-based instruments, the Kyoto mechanisms established:

- international emissions trading – which will allow developed countries to buy and sell emission credits among themselves;
- joint implementation – a way to earn emission reduction units by investing in emission reduction projects in other developed countries;
- The 'clean development mechanism' – a way to earn reduction credits by investing in emission reduction projects in developing countries.

The Kyoto Protocol was to enter into force when it had been ratified by fifty-five parties, including sufficient developed countries to represent at least 55 per cent of total carbon dioxide emissions in 1990. It came into effect in February 2005, when Russia finally having decided to ratify it. The main points of the Kyoto Protocol are summarised in Box 2.1.

Key developments after COP3
The main focus of this book is Japanese and British climate policy up to COP3; however, let me briefly look at milestones on international climate politics after COP3.

COP3 left some of the more difficult issues unresolved (see Box 2.1). At COP4, therefore, parties adopted the Buenos Aires Plan of Action, which established deadlines for finalising work on operational details of the Protocol, including rules governing the Kyoto mechanisms, compliance issues, financial assistance for developing countries and rules on crediting countries for carbon sinks. Parties, however, failed to reach consensus on a number of key issues at the deadline set of COP6. Principal stumbling blocks were the issues relating to sinks and 'supplementarity' in the use of the Kyoto mechanisms. The gap between the EU and G77/China, on the one hand, and the 'Umbrella Group' (the USA, Canada, Australia, New Zealand, Russia, Norway, Iceland and the Ukraine), on the other, remained wide.

After the initial collapse of COP6 there was a further blow to climate politics. In March 2001, the incoming US President, George W. Bush, announced that the USA would not implement the Kyoto Protocol, and described it as 'fatally flawed in fundamental ways'. Bush's announcement caused immediate criticism from within and without, and also had the world fearing that the Kyoto Protocol might not enter into force: the USA accounted for 36.1 per cent of carbon dioxide emissions among developed countries, and ratification of the Protocol by developed countries accounting for 55 per cent of the total 1990 carbon dioxide emissions was necessary for it to enter into force. Nevertheless, by the end of the month the EU environment ministers had agreed to pursue ratification without the USA, and the fate of the Kyoto Protocol was effectively in the hands of Japan and Russia, which together accounted for 25.9 per cent of 1990 carbon dioxide emissions among developed countries (see Table 2.4).

Box 2.1 *Main points of the Kyoto Protocol*

Quantified emissions reduction or limitation objectives
Gases covered: CO_2, CH_4, N_2O, HFCs, PFCs, SF_6.
Baseline year: 1990 for CO_2, CH_4 and N_2O; either 1990 or 1995 for HFCs, PFCs and SF_6.
Net or gross approach (sinks): Net approach, but emission removals counted are limited to afforestation, reforestation and deforestation that have taken place since 1990.
Target year: 2008–12 (the first commitment period).
Reduction/limitation objectives: Developed countries, including economies in transition (as specified in Annex B of the Protocol), shall reduce their overall emissions of the six key gases by at least 5 per cent below 1990 levels.

8% reduction: The EU (each member state of which has its own differentiated target), Bulgaria, the Czech Republic, Estonia, Latvia, Liechtenstein, Lithuania, Monaco, Romania, Slovakia, Slovenia, Switzerland.
7% reduction: The USA.
6% reduction: Canada, Hungary, Japan, Poland.
5% reduction: Croatia.
0% reduction: New Zealand, Russia, Ukraine.
1% increase: Norway.
8% increase: Australia.
10% increase: Iceland.

Banking and borrowing: The Protocol allows the banking of emission reductions below the required emission reductions in the first commitment period, but not borrowing.

Policies and measures
Developed countries are encouraged to work on the promotion of energy efficiency and renewable energy, enhancement of GHG

To entice Canada, Australia and, especially, Japan and Russia, at the resumed COP6 in Bonn in July 2001 and COP7 in Marrakech in November 2001 the EU and G77, with China, offered generous concessions on key issues relating to the use of sinks and the Kyoto mechanisms to attain the Kyoto targets. This move enabled the COP to reach agreement at the political level on detailed directions for key unsettled issues (the Bonn Agreement) sinks, reforms to tax/financial systems as necessary and reduction in GHG emissions in the transport sector and in methane emissions.

International emissions trading
Developed countries (as defined in Annex B of the Protocol) can trade emissions credits between themselves.
Major issues to be resolved: The definition of the relevant principles, modalities, rules and guidelines for emissions trading.

Joint implementation
A developed country (as defined in Annex I of the FCCC) can receive emission reduction credits if it funds a project in other developed countries where GHG emissions are actually reduced.
Major issues to be resolved: The elaboration of further guidelines on joint implementation.

Clean development mechanism
Developed countries (as defined in Annex I) can receive emission reductions credit by financing projects aimed at reducing GHG emissions in developing countries.
Major issues to be resolved: The elaboration of the modalities and procedures.

Assistance mechanisms for developing countries
Developed countries are urged to provide new and additional financial resources, including technology transfer, needed by developing countries to meet their existing commitments.
Major issues to be resolved: Inter alia, the establishment of funding, insurance and transfer of technology.

Compliance
Major issues to be resolved: The establishment of appropriate and effective procedures and mechanisms to determine and to address cases of non-compliance.

and, building on the Agreement, to adopt the Marrakech Accords at COP7. The Marrakech Accords worked out outstanding issues on rules of the Protocol, and thus provided the basis for countries' decisions on ratification of the Protocol. While paving the way for the Protocol's entering into force, however, it is said that 'loopholes' introduced at the resumed COP6 and COP7 reduced the real effect of the Kyoto Protocol from a 5.2 per cent cut in

Table 2.4 *Carbon dioxide emissions in 1990 among developed countries*

Countries	Proportion (%) of CO_2 emissions from all developed countries	Ratification of the Kyoto Protocol as of 5 July 2005
USA	36.1	No
EU	24.2	Yes
Russia	17.4	Yes
Japan	8.5	Yes
Canada	3.3	Yes
Australia	2.1	No
Norway	0.3	Yes
Switzerland	0.3	Yes
New Zealand	0.2	Yes
Central and eastern European countries (except for Russia)	7.6	Yes, except for Croatia

Source: http://unfccc.int (last accessed 11 July 2005).

GHG emissions in developed countries to a mere 1.5 per cent (*Environment Daily*, 12 November 2001).

After COP7 parties to the FCCC continued to negotiate, based on the Marrakech Accords, details of the rules for the Kyoto mechanism, in particular those concerning implementation of the clean development mechanism. With the Russian government's decision to ratify the Kyoto Protocol in November 2004, which allowed it to enter into effect as of 16 February 2005, the key contention is now 'post-Kyoto' arrangements for GHG emissions control that will involve the USA.

Implementing the FCCC[8]

The 'ultimate' objective of the FCCC is:

> stabilisation of greenhouse gas concentrations in the atmosphere at a level that would prevent dangerous anthropogenic interference with the climate system. Such a level should be achieved within a time-frame sufficient to allow ecosystems to adapt naturally to climate change, to ensure that food production is not threatened and to enable economic development to proceed in a sustainable manner. (Article 2)

This grand objective is pinned down by a more specific indicative target for the first step to be taken by Annex I countries (developed countries, including those with economies in transition). Article 4.2(a) requires them to take action in the recognition of the fact that:

> the return by the end of the present decade to earlier levels of anthropogenic emissions of carbon dioxide and other greenhouse gases not controlled by the Montreal Protocol would contribute to [modifying longer-term trends in anthropogenic emissions]....

Article 4.2(b) supplements this with 'with the aim of returning individually or jointly to their 1990 levels these anthropogenic emissions of carbon dioxide and other greenhouse gases not controlled by the Montreal Protocol' (see Table 2.1). The vague, ambiguous wording and the separation in the Article of the passages identifying the baseline year and the target year reflects controversy through the INC sessions and especially the USA's strong reservations about the inclusion of specific targets in the FCCC. These provisions are generally interpreted as meaning that developed countries were required to aim to return their GHG emissions to 1990 levels by 2000; however, countries were allowed to interpret this flexibly by taking into account their own economic circumstances, starting points and approaches.

Parties to the FCCC are obliged to adopt policies and measures to limit GHG emissions and to protect and enhance GHG sinks. In doing this, cost-effectiveness should be a guiding principle (Article 3.3). The Convention, however, does not specify what policies parties should adopt in order to attain the objective. Neither is there provision of sanction for non-compliance. The only instrument available to ensure that developed countries take appropriate action is the 'National Communication' which the FCCC obliges them to submit to the COP periodically. A National Communication is reviewed by a cross-national team but it does not make recommendations on specific policies and measures to be taken by a country.

The FCCC thus provides countries with significant latitude in its implementation. At the same time, however, it contains a potentially significant driving force for political, economic and social change. Article 3.3 adopts a precautionary principle: 'Where there are threats of serious or irreversible damage, lack of full scientific certainty should not be used as a reason for postponing such measures'. Article 3.1 is underpinned by the notion of inter-generational equity, and states that climate policy is 'for the benefit of present and future generations of humankind'. Intra-generational equity is also ensured under Article 3.1, which postulates the principle of 'common but differentiated responsibilities' and, implicitly acknowledging that the problem of global warming occurring at present has largely been caused by an extravagant use of energy in developed countries, requires them to take the lead (see Figures 2.6 and 2.7 for national contributions to carbon dioxide emissions). This is the reason why obligations

Figure 2.6 *The world's largest contributors to carbon dioxide emissions in 2000 (total 6.4 billion tonnes of carbon).*

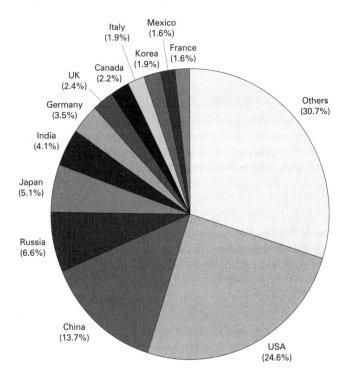

Source: The Energy Conservation Centre (2003).

Figure 2.7 *Carbon dioxide emissions per capita in selected countries and regions, 1990 and 2000. Note that the figures for Russia are for 1995 and 2000.*

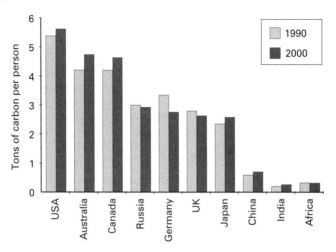

Source: Energy Conservation Centre (2003).

to submit a National Communication and the indicative target are specifically for developed countries. The FCCC also urges countries to review their domestic policy in light of the climate imperative. Article 4.2(e) requires countries to coordinate their administrative and economic measures if necessary, and review their own existing policies so as to integrate measures to control climate change, and 'the widest public participation' is also encouraged. The FCCC crucially requires countries to reorientate existing international relations, national policies and institutions towards a more sustainable society.

Conclusions

Despite remaining uncertainties surrounding global warming, as the scientific consensus on the human-induced greenhouse effect built up, the world began to act against this threat. On the crest of the global environmental wave in the early 1990s, more than 150 countries signed the FCCC, representing a collective commitment to the long road towards combating global warming.

In concluding this chapter, I would like briefly to consider the implications of the nature of the global warming problem for policy and political developments, since, as will be explained in Chapter 3, it involves theoretical issues. Attention should be paid to at least four points.

First, the problem is global. The atmospheric concentration of GHGs and their effects upon the earth do not respect national boundaries. As such, the fruits of a country's effort to reduce GHG emissions are not restricted to its territory but dispersed to other areas of the world. This is likely to discourage rational political actors from taking action. Moreover, countries are not equally vulnerable to the effects of climate change. Developing countries and the poor within the countries are more exposed to risks, largely because they have less capacity and resources to adapt to the threat, and the poor are often in the vulnerable locations. Thus, although the world shares the threat of global warming, the actual scale of disaster caused by climate change is expected to differ geographically, which means that any benefits accruing from tackling the problem are equally likely to differ internationally.

A global dimension is also involved in the causes of the problem as well as its effects. The most important GHG, carbon dioxide, is emitted by every country, none of which emits a decisive amount of the world total. Hence, the solution requires global cooperation. This, however, has to be sought in the context of wide variations in income levels, flexibility and expectations of the future (IPCC, 1996: para. 1.9). The problem of global warming poses an acid test for the capacity of the international community to develop meaningful cooperation.

Second, carbon dioxide emissions, mostly caused by the combustion of fossil fuels, are closely related to the socio-economic life of nations. Since the world's experience of the recession following the oil crises of the 1970s, reductions in carbon dioxide emissions implies economic slow-down. Significant reductions of these emissions also require a reorientation of existing socio-economic systems and, for industrialised countries, even business culture and lifestyles. In other words, the ultimate implication of a stringent carbon dioxide reduction policy is an expensive systemic overhaul. The difficulties facing policy-makers are obvious.

Third, the long time lag between the causes and effects of global warming necessitates extremely long-term perspectives. It is a disincentive for politicians to take action. What is important to politicians is making tangible achievements before the next election. Measures to cut carbon dioxide and other GHG emissions are unlikely to help them earn political points during their term of office because the benefits of reducing the risks of global warming will not be felt in such a time frame. Indeed, these measures may even decrease politicians' popularity, because of the costs entailed by a GHG reduction policy, the effect of which is, moreover, still uncertain.

Last, there still remain scientific uncertainties. They not only leave sceptics and policy-blockers room for contention, but also make uncertain the effects, especially at the regional level, of the immediate limitations on emission. Traditional cost–benefit policy analysis is consequently very difficult.

To bring the threads together, the problem is global; the level of emissions of the main GHG, carbon dioxide, is closely related to a country's welfare; there is a long time lag between causes and effects; and there still remain scientific uncertainties. These factors will present policy-makers with a formidable set of difficulties and complications.

Before proceeding to the next chapter, I should reiterate the focus of the study, that is, on politics and policy developments regarding limitations or reduction of national carbon dioxide emissions. Carbon dioxide is both scientifically identified and politically recognised as the most important GHG, and national emission levels have been the main concern among the international community, including Japan and Britain.

Notes

1   As explained in Chapter 3, in this book the term 'regime' is used as scholars of international relations use it.
2   The second assessment report, in particular, was contended by a handful scientists.
3   Recent scientific research has also raised concerns about the massive anthropogenic emissions of aerosols, most of which have local cooling effects, as well as reactive gases, which are not themselves GHGs but which influence the warming effect of GHGs through atmospheric chemical reactions.
4   There are, however, short-lived GHGs such as ozone, which consequently have spatial warming effects that differ from those of the

main GHGs. If these short-lived gases are included, ozone is the third most important GHG, according to the IPCC (2001a: paras 4.24, 6.31–34).

5 'No regrets' measures are those which, by definition, imply net benefits for the economy, whether or not the risk of climate change is real. Energy efficiency is such a measure. The case for no-regrets measures is often strengthened by additional positive external effects, such as reduced air pollution, or less traffic congestion, and so on. See OECD (1995: 8).

6 These targets were agreed not by government representatives but by individual attendees.

7 The Group of 77 (G77) at the United Nations is a loose coalition of developing countries established in 1967 with the aim of strengthening their joint negotiating position. Although it started with 77 member countries, it now has 133.

8 'Implementation' of the FCCC does not really mean 'policy implementation' in the conventional sense of political science, that is, the process by which prior policy decisions are translated into action. Because of the 'framework' character of the Convention, its implementation involves domestic policy formulation and developments. And it is this aspect of 'implementation' of the FCCC with which this book is concerned.

3
Frameworks of analysis: the institutional approach and the issue-based approach

As the subsequent chapters show, global warming policies in Japan and Britain exhibit both similarities and differences. Both governments were initially reluctant to take action; however, over time they changed their policy to meet international demands, and apparently took very different stances on the international stage. Although the contents of their policies and measures were different, the two countries behaved in a similar way in relation to certain aspects of policy. The questions to be asked are:

• How can we explain this mixture of policy similarities and differences?
• What are the important factors and how do they interact with each other in explaining this mixture?

As a way to approaching these questions, drawing on Freeman's contrast between the 'national style approach' and the 'policy sector approach' (Freeman, 1985), I will focus on two strands of the literature. One is (new) institutionalism and the other is the 'policy determines politics' school, or what I call the 'issue-based approach'. The institutional approach looks at different characters of national institutions, and predicts policy differences, while the issue-based approach looks at the constraints on rational actors arising from the attributes of a policy issue, and implies cross-national policy similarities. The combination of these two approaches, therefore, will give useful analytical frameworks within which to explain Japanese and British policies on global warming. International forces and the role of individual catalysts are also taken into account in doing this, but they are, in this study, essentially treated as exogenous stimuli to domestic policy processes.

The discussion of institutionalism is linked with the literature on corporatism, here understood as describing a form of institution that can be contrasted with non-corporatism, or pluralism. In this study, Japan is seen as having important institutional characteristics of corporatism *in terms of analysing environmental policy*, and Britain those of pluralism. As explained subsequently, it has been argued that the degree of corporatism has implications for environmental performance (see, for example, Dryzek, 1997: 140–1, 151–2; Lijphart, 1999: 293–8; Lafferty and Meadowcroft, 2000b: 424). The contrast between corporatism and pluralism, thus, provides a sub-framework for analysing global warming policies in Japan and Britain by approaching the secondary question of how institutions matter, if they matter.

The aspects of policy I am concerned with are: the speed of policy change; policy content; the degree of policy integration; and policy stringency. The *speed of policy change* refers to how long a country takes to change policy after an issue comes onto the national agenda. *Policy content* pertains to the general policy approach, including the choice of instruments. *Policy integration* is still an ambiguous notion. An analytical framework of policy integration was discussed by Lafferty and Hovden (2003: 12). They considered (environmental) policy integration to be mainly about 'strategies, plans, instruments or other initiatives' which are designed to promote a transition towards sustainable development and focused 'primarily on process and policy' as objects for analysis (p. 12). They then provided an analytical framework for policy integration which consisted of both vertical dimensions (sectoral integration) and horizontal dimensions (intra-ministerial integration).[1] In this study I will focus on the vertical dimension – that is, on how the energy and transport sectors have integrated considerations of global warming into their policy processes. Following Jänicke (1997: 13), I will also look at the 'external integration', which includes the integration of non-governmental actors. Finally, *policy stringency* covers governmental ambition in tackling a problem and its ability to pursue a necessary policy.

The structure of this chapter is as follows. First, I look at the institutional approach and the implications for the environment of corporatism and pluralism as institutions. Next, the issue-based approach and its implications for global warming politics are discussed. I then briefly explain international influences and the role of individuals. In the last section I outline the explanatory patterns the two approaches would provide in explaining the four aspects (set out above) of global warming policies in Japan and Britain.

The institutional approach

The role of institutions in politics has recently regained the growing attention of social scientists under the broad heading of 'new institutionalism' (March and Olsen, 1989). New institutionalists see institutions more as 'rules' than as 'organisations'. Institutions are not 'things' but 'processes' (Lowndes, 2002: 97–9). To borrow Hall's definition, they are 'the formal rules, compliance procedures, and standard operating practices that structure the relationship between individuals in various units of the polity and economy' (Hall, 1986: 19). Institutions thus include such things as a system of rules for electoral processes, a system constituting patterns of relations between government and interest groups, and regimes for international cooperation (see Peters, 1999).

The new institutionalism is, however, not a single approach, but 'one broad, if variegated, approach to politics' which is held together by the core argument that institutions are 'the central component of political life' and that 'institutions *do* matter' (Peters, 1999: 149–50, original emphasis). Among the approaches identified are 'rational choice institutionalism' and 'sociological institutionalism' (Hall and Taylor, 1996), but these are underpinned by rival ontologies and hence different theoretical assumptions, especially in relation to the interaction between individual actors and institutions. Nevertheless, it has been observed that the gap between the two approaches is narrowing through mutual recognition of the argument of the other (Lowndes, 2002: 106–7) and the potential of historical institutionalism in further developing the dialogue between the two (Thelen and Steinmo, 1992; Hay and Wincott, 1998; Hall and Taylor; 1998). Therefore, and also given the framework of analysis of this book, which opposes the institutional approach to the issue-based approach, the book treats the institutional approach broadly rather than choosing one from variety of different approaches. Nevertheless, given that historical institutionalism 'represent[s] a considerable advance' on its antecedents (Hay and Wincott, 1998: 954), the institutional approach in this book is largely in line with historical institutionalism.

How, then, do institutions influence politics and policy? Basically, institutionalists argue that institutions condition the behaviour of political actors. As such, institutions influence *strategies* that rational political actors adopt in trying to attain their interests. Various works on corporatism and macroeconomic performance (for example, Schmitter and Lehmbruch, 1979; Katzenstein, 1985; Pekkarinen *et al.*, 1992) and Weaver and Rockman's (1993) comparative analysis of government capabilities also suggest that by conditioning political processes in a distinctive way, institutions influence *governments' abilities* to perform specific tasks. Some strands of recent institutionalism further argue that institutions influence not only actors' policy strategy but also their *preferences* (Thelen and Steinmo, 1992). An early work of this strand is that of Katzenstein, who explained differences in objectives and instruments in foreign economic policy in advanced liberal democracies principally in terms of the structure of domestic institutional arrangements for policy processes (Katzenstein, 1978). Similarly, Hall argues that differences in the organisation of economic policy-making in Britain and France affected both the perceptions of what is rational and the distribution of power, and hence generated different policy outcomes (Hall, 1986).

One distinctive view of institutions that has recently attracted students of new institutionalism relates to the way in which institutions interact with ideas (Hall, 1989; King, 1992). Institutions are products of past political struggle; as such, they encapsulate winning interests and ideas at the time of institutional formation or change, which occurs when the existing institution fatally fails to serve an actual demand (Krasner, 1984; Thelen and Steinmo, 1992; Hall, 1993). At the same time, institutions are autonomous and reproduce themselves (Krasner, 1984). Institutions thus function as *mobilisers of bias*. They generate vested interests, and mobilise and magnify affinitive ideas and interests at the expense of other ideas in the way that interests of the dominant actors are served (Thelen and Steinmo, 1992; Hall, 1992). In this process, certain ideas are institutionalised, and hence they become inseparable from the institutions themselves. From this perspective, institutions are broadly defined to include the ideas they incorporate.

An important implication here is that in well established, cohesive institutions, Lukes' second and third dimensions of power are in operation (Lukes, 1974: 36–45). That is, power is exercised not only through observable actions (the first dimension of power) but also through a conscious effort to keep certain issues off the agenda (the second dimension) and through social construction (the third dimension). This means that for fundamental changes in policy to occur, the agent of policy change has to engage in a 'war of position': it needs first to change the prevailing 'world views', 'cognitive maps' and beliefs before its ideas become more acceptable (Smith, 1993: 93).

From these institutionalist arguments there follows a particular view of the role of ideas in understanding public policy. The conceptual underpinning of this view involves belief systems. In a simple form proposed by Majone (1989: 150–4), belief systems consist of a core and peripheries.[2] The core is the stable, rigid part of policy, which is surrounded by several concentric, flexible peripheries (which construct 'protective belts'). Policy-makers try to translate their core beliefs into policy and to defend them by changing the policy peripheries when challenged. The core point of policy-making in this view is the ongoing search for, and adaptation of, policies through learning (Hall, 1993: 278; Jenkins-Smith and Sabatier, 1993: 44) and it is through discussion and argument that what is rational is defined and preferences are shaped (Majone, 1989).

Institutionalists argue, however, that these intellectual aspects of the policy process should be considered in conjunction with institutional conditioning. Ideas need institutional vessels in which to reach the locus of policy-making (Majone, 1989: 161; Hall, 1986: 280). Also, the selection of an idea is conditioned by institutional biases as well as institutional opportunities for and constraints on learning.

Corporatism and pluralism as contrasting institutions
Corporatism[3] is generally understood as interest intermediation and a pattern of policy processes, especially in macroeconomic management. Recently corporatism has been applied to new, 'postindustrial issues', such as the environment (Lijphart, 1999: 174).

Under corporatism, the main societal actors, business and labour, are well organised into encompassing, hierarchically ordered peak associations. The state engages in collective policy-making with these associations. Typically they have conflicting interests in the short term, but engage in collaborative bargaining in search for zero-sum or even positive-sum outcomes in the medium to long term. Here, the state takes an interventionist role, engaging in strategic economic planning – an attribute of corporatism. Responsibility for policy implementation is shared between the key players. Through such cooperative bargaining at the top and the mobilisation of compliance through the peak associations, a high level of policy concertation develops (Schmitter and Lehmbruch, 1979; Panitch, 1980; Lehmbruch and Schmitter, 1982; Williamson, 1989: 102; Schmitter, 1989). These characters are an ideal type and liberal democracies almost always fall somewhere between pure corporatism and pure pluralism or non-corporatism (see below). Countries with strong features of corporatism include Austria, Norway and Sweden.

Corporatism's style of policy-making, based on consensus and cooperation, is shared by the consensus model, developed by Lijphart, of government forms (see discussions by Lehmbruch and Schmitter, 1982; Lijphard, 1984; Lijphart and Crepaz, 1991; Crepaz and Lijphart, 1995; Keman and Pennings, 1995). Consensus democracy aims to obtain broad agreement on policy that the government pursues. Power is shared in the form, for example, of a coalition government, and the concertation of partisan interests takes place. An implication is that both interest group politics (corporatism) and legislative politics (consensus democracy) are governed by the same norm of collaboration and concertation. Although the two systems are not identical, there is empirical evidence of correlation between them (Crepaz and Lijphart, 1995).

Turning to pluralism, I follow Lijphart in characterising it as an institution as 'non-corporatism', with institutional characteristics that contrast with those of corporatism (Lijphart, 1984). Thus, in contrast to the stable set of a limited and select number of groups in the corporatist policy-making arena, that of pluralism is open and replete with multiple, decentralised and overlapping groups (Dahl, 1961; Dunleavy and O'Leary, 1987; Smith, 1995). Groups 'combine, break, federate and form coalitions' in pursuit of their self-interests, and constellations of power are 'in [a] flux of restless alteration' (Latham, 1964: 49). Even if institutionalisation of relationships is observed, the emphasis remains on groups' capability to break it (Jordan and Richardson, 1987: 278–91). Hand in hand with this group dynamism is the pattern of the policy process characterised as 'pressure' (Schmitter, 1982: 263). The government accords more influence to the most vocal groups, and this generates a policy-making process that approaches 'winner takes all' (Dunleavy and O'Leary, 1987: 35; Hill, 1997: 33). The policy process is competitive and conflictual from the outset to the end. Policy concertation is alien to pluralism, and government and private interests operate essentially in separate worlds (Schmitter, 1982: 263; Smith, 1995: 209, 211; Crepaz, 1995: 392, 396–7). An example of non-corporatism or pluralism is the USA, where there is no peak association capable of rounding up diverse interests in business and labour, nor a tradition of state intervention, which discourages both the state and groups from forming a system of interest intermediation (Wilson, 1982).

In the same way that corporatism is associated with consensus democracy, pluralism has a conceptual affinity with the majoritarian model of legislative democracy. Lijphart's concept of majoritarian democracy, modelled on the Westminster parliamentary system, depicts a government by the majority of the people: the party that wins the majority of seats in the parliament has decisive control over policy formulation. Pluralism and the majoritarian democracy share the decision-making style of majoritarianism and 'winner takes all'.

The patterns of policy processes that consensus corporatism and majoritarian pluralism are associated with can be elaborated by building on the concept of policy styles (Richardson *et al.*, 1982). According to this concept, a society develops 'standard operating procedures' and 'legitimising norms' for formulating and implementing public policy. National policy styles follow prevailing forms, which Richardson *et al.* (1982) suggest are nationally idiosyncratic.[4] Policy styles are, in a sense, an expression of combinations of various institutional predilections for particular ways of processing policy. In this sense, the national style approach is a specific version of general institutionalism. In identifying policy styles, Richardson *et al.* resort to two primary features of the policy process, that is, the government's approach to problem-solving (proactive or reactive) and the relationship between the government and other actors in the policy process (consensual or impositional).

According to this conceptualisation, pluralism – which is characterised by the winner-takes-all decision-making style, the 'pressure'-based political process and the absence of the practice of state intervention – predisposes to reactive, impositional styles. On the other hand, the institutional characteristics of corporatism certainly suggest consensual policy styles, while strategic economic planning implies an institutional capacity for proactive policy styles.

An important contrast between corporatism and pluralism concerns the type of policy network that each is associated with. The term 'policy network' is used to designate sectorised policy-making and implementation (Rhodes, 1988; Marsh and Rhodes, 1992; Wilks and Wright, 1987).[5] An analysis of policy networks sheds light on the characters of and relationships between the state and non-state actors who are involved in processing policy in discrete issue areas. In a tight policy network, the actors are in strong resource/power dependency relationships and their relationships are institutionalised. Implicitly, a tight policy network provides the government with the capabilities to intervene in and draw cooperation from societal actors for effective policy implementation. In Smith's terms, the state has strong infrastructural power (Smith, 1993). In a loose policy network, on the other hand, relations among actors are weak. Power dependency relations and infrastructural power are negligible. We can see the logical implications of corporatism and pluralism for the type of policy networks. With corporatism, where reciprocal relations develop between the state and the market, policy networks are likely to be tighter and more closed than is the case with pluralism.

Institutional implications of corporatism and pluralism for the environment

The advantages of corporatism over pluralism

There are at least three reasons why corporatism can be better for the environment. First, 'inherently intrusive' environmental regulations are more acceptable and feasible in a country where government intervention has greater legitimacy (Scruggs, 1999: 5).

With corporatism, the government intervenes in markets and, importantly, asserts a strategic policy orientation. According to Jänicke (1995: 15; see also Jordan and O'Riordan, 1993: 187), strategic orientation is 'the highest stage of institutional capacity' for successful environmental policy, as the ultimate objective of environmental policy is to change economic behaviours and structures into more environmentally acceptable ones. Moreover, the tradition of intervention and market-steering provides institutional conditions on which the view flourishes that environmental regulations create new market opportunities for environmental technologies. This is consistent with corporatism's institutional capacity to transform essentially zero-sum relations into positive-sum relations. Corporatism may well change the definition of what the 'interest' of environmental policy should be.

Second, cooperative relations between the regulator and the regulated are likely to help enhance the effectiveness of government policy. Industry may be more ready to share important information in setting environmental standards and to act on them than is the case where there are combative relations (Wallace, 1995).

Third, corporatism's encompassing groups indicate that there is the institutional capacity and incentive to mobilise collective efforts in pursuit of public goods. Policy compliance can be secured or facilitated through discipline and side payments to member groups. Member groups, much less concerned about free-riders, also have a greater incentive to comply with policy.

To bring the threads together, corporatism has the institutional capacity to pursue the public good of environmental protection strategically and in a concerted manner. Compare these implications of corporatism with those of pluralism. Decentralised, competitive business organisation provides little incentive to cooperate in pursuit of public goods. Pluralist non-interventionism would also make the government hesitate to take the necessary action, and even if taken may lack strategic orientation. Moreover, in a pluralism characterised by 'pressure' politics, where 'offence, or at least "active resistance" is the best defence' (Latham, 1964: 55), a policy is more prone to implementation failure. Its reactive policy styles may also encourage the regulated industries to continue resisting stringent environmental policies. The comparative advantage of corporatism has been empirically demonstrated by Scruggs (1999, 2001).

The advantages of pluralism over corporatism

The argument for the superiority of pluralism over corporatism is based on pluralism's institutional advantage in establishing a prompt, rich and deep policy debate. Institutions enable as well as constrain. With corporatism, the policy-making process is exclusive as much as inclusive because it accords a limited number of select interests privileged access to and status in the elitist policy-making process, while others are excluded from it. Consumers and environmental interests are often among those excluded. Opschoor and Straaten (1993), for example, argue that in the Netherlands the strong political power of capital and labour pushes out 'the interests of powerless natural resources'. Moreover, institutions mobilise biases. The interests of capital and labour, if incongruent with the interests of the environment, can be very resistant to the latter. Indeed, typical corporatism, which pursues economic growth, may well be entrenched in a 'productivist ideology'.

The advantage of open, less institutionalised pluralism is clear. In a pluralist system, where no source of power is dominant, once an environmental voice reaches a sufficiently high 'decibel rating', the government responds to it quickly. A related issue is the ease with which pluralism allows environmentally damaging policies to be abolished. With corporatism, to the extent that policy is based on a consensus between interested actors whose relations are institutionalised, reversing policy entails high political costs. With pluralism, policy is likely to be less entrenched.

The inclusion of environmental interests in corporatist policy processes would greatly facilitate environmental policy. Nevertheless, a pitfall is that the incorporation of environmental concerns may be, in fact, a strategic action to defend the existing consensus on the primacy of economic growth: McEachern (1993) shows how a corporatist inclusive policy process can co-opt or assimilate 'socially disruptive elements of environmental concern' into the existing political discourse, thus deflecting the immediate environmental pressures while defending the existing consensus on economic development.[6] Hukkinen similarly argues that in Finland conflicting environmental interests are systematically integrated into corporatist policy-making institutions, whose design inevitably 'prevents the long-term viability of eco-systems from becoming the central guideline for decision-making' (Hukkinen, 1995a). With corporatism, therefore, there is a risk that fundamental and long-term environmental questions become obscured in the search for immediate consensus, thus making them secondary to shorter-term politico-economic expediencies. Because prioritising economic interests is the consequence of institutionally rational behaviour, Hukkinen (1995a,b) argues that corporatism has inherent institutional impediments to long-term environmental protection. The practice of consensus-building through mutual accommodation and concession may well result in second-best solutions.

Again, pluralism compares favourably in this respect. Its open, competitive political arena allows various conflicting views and interests to emerge in the political process. Conflict is articulated and not obscured. Moreover, unlike corporatism, where decisions tend to be made through political exchange for short-term expediency among a few elites, pluralism's open political arena encourages the discussion of environmental issues. Pluralism can, therefore, provide a good politico-institutional framework for the evolution of green ideas (De Shalit, 1995).

The issue-based approach

The argument so far is based on the institutionalist claim that institutions have distinctive influences on politics and policy. There are, however, at least two reasons why the institutionalist approach is complemented by a discussion of issue characteristics. First, whether institutions work more to constrain or to provide opportunities may well depend on issue characteristics, as implied by 'mobilisation of bias' (Schattschneider, 1960: 71). Second, as Hall and Taylor (1996: 939) point out, 'it is through the actions of individuals that institutions have an effect on political outcomes', and individuals who process an issue are fundamentally conditioned in their actions by that issue understood as a particular configuration of constraints. By making use of the issue-based approach in our analysis, we can better predict the relationships between institutions and policy. However, what the issue-based approach implies in terms of comparative policy analysis is the opposite to that of the institutionalist approach. While the logical corollary of the latter is cross-national difference, the former indicates cross-national similarity.

The issue-based approach starts with the idea that the nature of an issue imposes constraints on policy options. From this follows two distinctive arguments regarding politics and policy. First, rational policy-makers looking for the best strategy will make a similar policy choice. This tendency is reinforced by cross-national policy learning and emulation, as well as by the influence of an international epistemic community (see below) in disseminating certain professional policy ideas into domestic policy formulation. Here, unlike the institutional approach, which highlights institutional conditioning of ideas, the issue-based approach assumes some extent of intellectual autonomy. At the same time, however, while the institutional approach allows scope for the intellectual and cognitive aspects to (re)define a problem (and hence interests and preferences), these aspects are screened out under the issue-based approach, since they are treated as given. In short, under the issue-based approach ideas can be instruments for attaining a 'given' desired goal but not a force for altering them (Blyth, 2002: 306).

The second and more theory-oriented argument is that there are only limited types of policies, which are, in turn, connected to the kind of politics and its outcomes. The most well known model of this thought is presented by Lowi (1964). According to him, there are basically only three types of policies – distributive, regulatory and redistributive (see Table 3.2 below) – and each type generates a predictable pattern of politics by evoking certain political interactions.

Another major school of thought concerning policy typology stems from public choice theory, which analyses how rational agents behave in the contexts of collective action. The central argument of this theory is that office-seeking politicians, and bureaucrats who try to maximise their organisational budget in expectation of rises in their salary and status, will produce a policy that favours business at the expense of the wider public, because of the resources that business possesses (e.g. financial support for electoral campaigning and lucrative positions after retirement), as well as the greater effectiveness with which business lobbies government.

The argument of the different degrees of effectiveness of organisational activities is based on Olson's logic of collective action (Olson, 1965, 1982). According to Olson, the existence of a common interest among a group of people does not necessarily lead to collective action; it depends on the size of the group in question. Let us take the example of the problem of pollution. The public may think about pressuring government to put a pollution-abatement policy in place. However, such a campaign entails time, energy and money. Because each individual's contribution to such collective action is very small, rational individuals will be tempted to free-ride the contributions of others. But this undermines collective action or, in this case, the campaign. The worst case would be no campaign, hence no desired policy, as a result of everyone trying to free-ride each other. The bigger the number of people affected by pollution, the less incentive an individual will have to engage in collective action, because his or her contribution matters less. In the politics of pollution, this leads to the public suffering from it. The many members of the public have difficulty in taking collective action, while the limited number of polluting factories have a greater incentive to engage in a campaign against strict environmental regulations, which would be costly for them. The corollary is policies that tend to serve smaller groups like business at the expense of the public. Moreover, the ordinary public will be 'rationally ignorant' of details of government policy. Rational individuals will think of free-riding others' effort to acquire information and expertise on policy affecting the public at large.

Wilson (1972, 1980b) provides an account of how policy can determine politics based on this logic of collective action. He classifies policy issues according to the incidence of costs and benefits. The underpinning idea is that the distribution of interests in a policy issue affects incentives for collective action, and hence the nature of group activity, thus shaping the pattern of politics.

Table 3.1 shows four possible situations involving concentrated or diffused costs and benefits. In cell 1, a policy proposal benefits a well defined small group but at a cost to another small group. Each side has a strong incentive to engage in group activities. The

Table 3.1 *Four possible situations involving concentrated or diffuse costs and benefits of a policy issue*

	Costs	
Benefits	Concentrated	Diffuse
Concentrated	Cell 1	Cell 2
Diffuse	Cell 4	Cell 3

Source: Self (1993: 29).

political outcome will reflect the balance of power between the two groups (interest group politics). In cell 2, where a policy proposal benefits a small group but the cost is diffused among a wide public, the beneficiaries will organise themselves effectively and lobby the government successfully, while rational individuals in the large group will fail to act for their group interest (client politics). 'Pork barrel' programmes are examples. Cell 3 is the case in which a policy generates diffused costs and benefits. Neither group will be incentivised to organise their activities. These policy proposals need popular support in order to be adopted (majoritarian politics). Welfare programmes are examples. Last, in cell 4, a policy imposes the cost on a specific, identifiable group for the benefit of a general constituency. The small group on which costs fall will organise effectively to block the policy or to reduce the burden on them, while the large number of beneficiaries will suffer from the problem of collective action. Typically, policies in this category are difficult to get adopted; however, skilled policy entrepreneurs may successfully push the policies through (entrepreneurial politics).

Wilson's policy typology is clearly different from Lowi's. Nevertheless, the patterns of political process that Wilson envisages have an obvious affinity to those envisaged by Lowi, as shown in Table 3.2. Entrepreneurial politics as described under Wilson's policy typology, however, sits uneasily with Lowi's policy typology. As Wilson points out, this is a new variant of a broad range of 'regulatory policies' since, unlike Lowi's regulatory policies, Wilson's entrepreneurial politics denotes competition between groups with unbalanced organisational capacity (Wilson, 1972: 329–30). It is this 'entrepreneurial politics' that global warming policy is chiefly about. The need to reduce carbon dioxide emissions means that those who produce and use fossil fuels intensively will be hit hardest. On the other hand, the benefits (avoidance of the disastrous consequences of global warming) are widely distributed, essentially to future generations.

The argument that a policy issue conditions patterns of political interactions is directly linked to the concept of policy networks, because policy networks are about interactions and relationships between actors who share interests in an issue. From the institutionalist perspective, I have considered the types of policy network that corporatism and pluralism are associated with. From the

Table 3.2 *Comparison of Lowi's and Wilson's policy typologies*

Lowi	Examples	Wilson
Distributive policies: patronage politics; a 'log-rolling' coalition (i.e. one in which groups with unrelated interests trade political support with each other)	'Pork barrel' programmes	Client politics: concentrated benefits and diffused costs
Redistributive policies: competition among broad categories of citizens; coalitions based on ideological ideas as much as interests	Income tax, welfare state programmes	Majoritarian politics: diffused costs and benefits
Regulatory policies: conflict between particular sectors or segments of the economy; competing coalitions based on common interests	The maintenance of fair-trade laws	Interest group politics: concentrated benefits and concentrated costs
–	Environmental regulations on factories	Entrepreneurial politics: concentrated costs and diffused benefits

issue-based approach, different types of policy network have different implications for collective action over an issue, because a policy network is seen as a product of collective action by those whose interests are at stake. Thus, if an issue involves more 'concentrated' interests (i.e. those involving smaller and/or fewer groups), the tighter a policy network will be.

Issue characteristics
So far I have looked at how the nature of an issue can be associated with a certain pattern of politics. The specific attributes of an issue, whatever the institutions within which that issue is processed, also affect policy developments. I have set out in Chapter 2 the characteristics of the problem of global warming and their implications for policy developments. Here I add three things from theoretical perspectives.

First, the problem of global warming is systemic and, at present, there is no practicable technology with which to capture carbon dioxide, the main cause of the problem. According to Jänicke (1997: 9), a government's capacity to cope with environmental problems is affected by whether the problem is systemic or structural in nature, how economically important the activities which generate the problem are, and whether a technical or other solution is available. In this sense, government's capacity to act is likely to be circumvented by the characteristics of global warming.

Second, across the global commons, the problem of global warming is that of a common-sink resource. Unlike common-pool resources, such as fish or minerals, common-sink resources – like the atmosphere, transboundary rivers and the oceans – cannot be appropriated. There are some reasons to argue that this characteristic of common-sink resources makes tackling the associated environmental problems more difficult. One is that while the burdens of over-exploitation of common-pool resources primarily fall on the exploiters, the damage caused by over-exploitation of common-sink resources would easily spill over on to actors other than the exploiters, or even be transferred to different actors. For example, over-fishing chiefly hits those over-fishing. However, in the case of the acid rain problem, the costs of sulphur dioxide emissions by Britain (an instance of the exploitation of the atmosphere) fell largely on Nordic countries. In the case of global warming, there is a gap between the level of contribution that a country has made in creating the problem and the level of damage that country is likely to suffer from the problem. This makes collective (global) action more difficult and complicated.

Also important is an issue's ability to attract public attention. Public attention and pressure are an important source of government incentive to tackle the problem and its capacity to carry out policies which may be unpopular among a small segment of society, such as a particular industry. Solesbury argues that people will tend to feel more strongly about issues with greater particularity and visibility (such as catastrophic events) than about those with cumulative effects over a relatively long time, and about those issues where the cause–effect links are relatively simple and obvious, and where responsibility can be clearly laid (Solesbury, 1976; Robinson, 1992: 105). The complicated cause–effect relations, scientific uncertainty and the cumulative effects of global warming make the necessary mobilisation of public support difficult.

The overall implications of the issue-based approach for global warming policy are straightforward: policy-makers face tremendous political and technical difficulties in developing policy, regardless of national institutions. The issue-based approach emphasises that the same constraints posed by an issue have crucial effects on different policy-makers' ability to develop robust policy.

International influences

Although states remain sovereign, as the world becomes increasingly interdependent we cannot afford to underestimate the extra-territorial influence in understanding public policy (Rose, 1991: 460–1). Of course, the causal relations are not one way: international politics has domestic sources as much as domestic politics has international sources. For the purpose of analysis, however, this study treats international influences simply as an exogenous stimulus to the domestic policy process. The aim of this study is largely to examine how two different countries have responded to the international demand to tackle global warming, and to explain their patterns of behaviour. So long as my concern is confined to the effect of international forces on the domestic policy process, to see international forces as a given does not substantially distort my analysis. The sources of international stimuli with which this study is chiefly concerned are epistemic communities and international regimes.

Epistemic communities are defined by Haas (1990: 349) as 'transnational networks of knowledge based communities that are both politically empowered through their claims to exercise authoritative knowledge and motivated by shared causal and principled beliefs'. In the case of global warming politics, the IPCC is an epistemic community. Haas's work on Mediterranean environmental protection demonstrates how an epistemic community can influence international and domestic policy by changing the attitudes of the national policy-makers; in his example coordinated action was taken to save the Mediterranean, after common scientific ground had been established for constructive international negotiation (Haas, 1989). Haas, however, does not claim an equal influence of epistemic communities for every issue.

Their authority and influence are likely to be conditioned by the economic and political implications of an issue. Indeed, as far as global warming is concerned, Haas speculates that epistemic consensus is unlikely to steer successful collective action and policy developments (Haas, 1990).

International regimes have been perhaps the most important exogenous stimulus in transnational environmental policy. An international regime is a form of institution (Young, 1989: 29), and includes international treaties as well as international organisations. The international regimes of global warming include the United Nations FCCC and the Kyoto Protocol. As an international institution, the regime frames the courses of action of nations and has independent consequences for domestic politics and policy. One clear example of such a consequence is the way the regime helps reduce the susceptibility of an issue to the 'ups and downs of the issue attention cycle' (Downs, 1972). The regime maintains or even increases concern over the issue on the part of policy-makers, pressure groups and the public (Krasner, 1983).

How, then, do these international forces affect the domestic policy process? After all, the final decision of whether to shift the national stance as demanded within the international sphere is in the hands of national authorities.[7] There are three possible answers to this question.

First, rewards and punishments offered by regimes, international pressures from other countries, and policy learning through an epistemic community may drive actors to recalculate costs and benefits, leading to changes in preferences. Second, international forces may alter the domestic political balance of power, by giving a 'tail-wind' to the minority and the undecided in the uncertain international political environment (Putnam, 1988). More specifically, Schoppa (1993) argues that one of the main effects of international pressure is participation expansion. Internationalisation of an issue may politicise it and mobilise public opinion, leading to an opening up of a hitherto closed, privileged policy network. International regimes can also directly affect the network by requiring new institutions to be established to deal with the regime objectives or demands (Haas, 1989, 1992; Keohane *et al.*, 1993). The new actors in the emerging looser network may include resourceful proponents of international demands, who had been placed on the periphery of, or excluded from, the previous policy network. This will entail changes in the power balance and may lead to changes in policy. Third, the international policy process can expand policy options beyond unique national constraints. 'Synergistic linkages' is the term Putnam (1988) uses to describe the situation in which issue linkages at the level of international negotiation expand the domestically feasible policy alternatives. Moreover, international demands may offer specific policy solutions to problems which may have not been considered earlier, owing, for example, to national institutional constraints (Schoppa, 1993).

International forces, thus, may well stimulate a new phase of political developments. Moreover, what is implicit is that international forces may reduce the effect of domestic institutions on politics and its outcome. For example, the tighter policy networks associated with corporatism may be opened up to become looser, more pluralistic networks.

Individuals as policy catalysts

Another important stimulus to the policy process that this study looks at is individual political actors. Among the major work discussing the role of individuals is that of Kingdon (1984). He argues that an individual can couple a problem and his or her pet policy idea at the time when the political climate is favourable for this, thus setting a new agenda item and paving the way for policy change. Politically powerful individuals, in particular, can directly influence the agenda and hence policy developments. A related issue here is the role of individuals in giving institutional force to ideas in the political arena. Institutionally resourceful individuals can function as 'transmitters' and 'shapers' of ideas (Dudley and Richardson, 2000: 238).

Even if an individual is not politically very powerful, he or she can still have role to play by building a winning coalition. Also important in policy developments is the role of the policy entrepreneur, as suggested in Wilson's policy typology (see above). Ralph Nader and Senator Edmund Muskie are well known US examples of policy entrepreneurs, acting for a wide public over the problems of automobile safety and emissions, respectively, in the 1960s. The efforts of individuals to catalyse policy developments can be helped by crisis (Wilson, 1980: 370–2). Crisis can effectively provoke the public's sentiments, on which a skilled individual can capitalise in the process of mobilisation. The effectiveness of individuals also heavily depends on the attitudes of third parties, including the media, commentators and political activists (Wilson, 1980: 371). They can help raise the public's sympathy for the cause for which individual catalysts are acting.

Framework of analysis

This book posits the institutional approach and the issue-based approach as two theoretical architectures for explaining public policy, and international influences and individual policy catalysts are treated essentially as stimuli to issue processing of the kind envisaged under the two approaches. What, then, are the implications of this framework for Japanese and British policy on global warming?

Before exploring this question, I would first like to make clear that in this study the terms 'corporatism' and 'pluralism' are used to denote important institutional attributes *in terms of analysing environmental policy and politics*. Specifically, corporatist institutional features relevant to environmental policy analysis – such as consensus policy styles, concerted action, interventionism, a relatively well integrated business sector (as reflected in the presence of numerous and effective business associations, including peak associations) and inclusive as well as exclusive policy processes – are collectively termed 'corporatism', while pluralist institutional features relevant to environmental policy analysis – including majoritarianism, non-interventionism, a fragmented business sector and a relatively open, competitive political arena – are collectively termed 'pluralism'. 'Corporatism' and 'pluralism' in this work are, therefore, not used as conceptualisations of the political system as a whole, but as a *heuristic shorthand* for two different sets of institutional attributes.

These institutional characters of Japan and Britain will be looked at in the next chapter. Nevertheless, I briefly look here at the empirical evidence on the levels of consensus corporatism in Japan and Britain. Figures 3.1 and 3.2 show the levels of corporatism in thirty-six democracies and their types of Cabinets and party systems, which can roughly be seen as representing the levels of consensus democracy.

Figure 3.1 *The relationship between type of Cabinet and interest group pluralism, 1945–96. The percentage of one-party Cabinets is the proportion over the period of Cabinets in which there was a one-party majority. Interest group pluralism is a rating derived by Lijphart (1999), partly on the basis of Siaroff's (1998) work.*

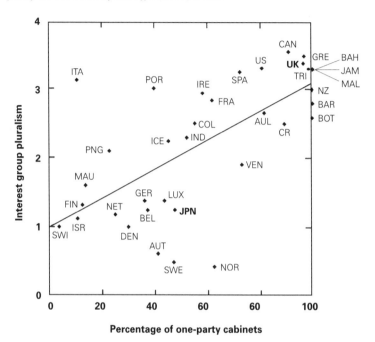

Source: Lijphart (1999: 182).

Now that it is clear what this study means by 'corporatism' and 'pluralism', let me briefly explain why Japan is treated as corporatism and Britain non-corporatism or pluralism in this study (detailed explanation is made in Chapter 4).

As shown, Britain scores high on levels of majoritarian pluralism. Since 1997 there have been moves away from the majoritarian system, such as the employment of proportional representation in elections to the devolved parliament and assemblies of Scotland, Wales and London. However, even if Britain has become a less centralised state, the majoritarian system in legislative politics clearly predominates, not least because of the limited power devolved (Dunleavy *et al.*, 2003: 9–12).

Figure 3.2 *The relationship between the effective number of parliamentary parties and interest group pluralism, 1945–96.*

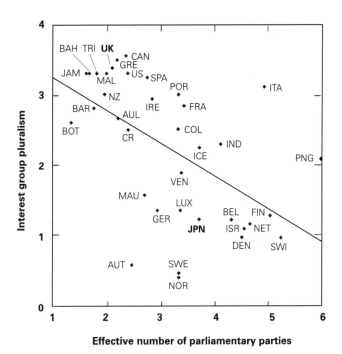

Source: Lijphart (1999: 183).

Japan, on the other hand, does not rate as highly on consensus corporatism as Britain does on majoritarian pluralism. When corporatist analysis became popular among social scientists, the question of whether Japan should be seen as corporatist was a controversial one. The main reason for this was that although labour is incorporated in firm-level decision-making, it is systematically excluded at the national level. More recently, scholars of Japanese politics have tended to emphasise the pluralistic and adversarial tendencies of Japanese politics (see, for example, Muramatsu and Krauss, 1987; Calder, 1988; Schwartz, 1998). It is certainly true that business is less integrated than before (again, in terms of the predominance of business associations) and that pluralistic politics has emerged as policy-making has become increasingly sectoralised. However, the point is that Japan can be still posited to be on the consensus corporatism side in relative terms. Moreover, there appears to have been a shift towards consensus democracy. This can be seen in the recent introduction of a parallel plurality proportional representation system in elections for the House of Representatives and the formation of coalition governments, especially since 1993. Also, to reiterate, this study is concerned with the institutional characteristics associated with corporatism, and not the Japanese political model or system *per se*.

Implications of the two approaches for Japanese and British policies on global warming

At the beginning of this chapter, I made clear that the focus of my policy analysis is: the speed of policy change; policy contents, including the choice of instruments; the degree of policy integration; and the stringency of policy. Now that the treatment of the Japanese and British examples has been made clear, we can suggest the types of policies on global warming that are likely to be countenanced by consensus corporatist and majoritarian pluralist states, and then consider the implications for Japanese and British policies.

The speed of policy change

Tsebelis's thesis on the influence of 'veto players' (i.e. those whose agreement is necessary to produce policy changes) on policy stability and change gives an insight into how quickly policy can change under consensus corporatism and majoritarian pluralism. According to Tsebelis (1995), policy-making institutions that embrace fewer veto players have a greater capacity to change policy than those with more veto players. Consensus corporatist institutions that involve multiple decision points are, therefore, likely to change policy more slowly than majoritarian pluralist institutions. The pluralist decision style, based on 'winner takes all', also suggests that, in pluralist systems, policy changes quickly once a majority is formed or after the power balance among political actors changes.

The issue-based approach, on the other hand, emphasises the influence of issue characteristics on the speed of policy change. In the case of global warming, policy developments are not easy to achieve, and hence policy change is likely to be slow.

Policy contents and instruments

The basic argument of the issue-based approach is that an issue constrains policy options. Rational policy-makers, then, choose the best strategy for a particular problem. The result is policy convergence. A more useful insight has been put forth by both Bressers and O'Toole (1998) and Daugbjerg (1998), who consider the influence of the nature of the target of a policy on the choice of policy instruments. When a policy targets diffused groups, it is practically impossible to apply instruments that have been tailored to a specific set of conditions. Policy instruments are, therefore, likely to be universal. When target groups oppose a policy, those instruments that give the groups freedom to decide whether or not to apply them, such as informative instruments, are unlikely to be used without any normative instruments, penalties or negotiated agreement. When a policy benefits the target groups, not only is compliance likely to be made voluntary, but heavy-handed regulations and penalties on failure to comply will be unnecessary. On global warming policy, we are likely to see universal policy instruments for carbon dioxide emissions by individuals and instruments with some compulsory elements, especially for energy-intensive industries.

The argument for the institutional relevance of the instrument chosen is put forth by Majone (1989) and Howlett (1991). Since it is within institutions that a policy is formulated, the cognitive maps associated with the institutions are likely to guide policy. From this perspective, we are likely to see distinctive institutional traits in policy contents and the choice of instruments.

It is possible to identify some implications of corporatism or pluralism for policy contents. First, a corporatist policy is likely to be more interventionist and oriented towards steering the economy in a desired direction rather than constraining certain activities. Second, and coupled with the first point, the close, collaborative state–industry relations under corporatism are more likely, than those under pluralism, to make policy-makers turn to subsidies and other kinds of financial support. Third, self-regulation, rather than heavy-handed coercive regulation, is likely to be used more often under corporatism than under pluralism. Last, Bressers and O'Toole (1998: 229–30) argue that if good communication channels are available to policy-makers, policy implementation under corporatism is more likely to be carried out directly by policy-makers rather than to be entrusted to other organisations.

The degree of policy integration

The degree of policy integration can be considered in terms of the tightness of the existing policy networks with which global warming concerns are to be integrated. From the perspective of the issue-based approach, the tightness of policy networks of the same sector across countries is likely to be similar, since a policy network can be seen as an effect of collective action by actors with stakes in the policy. Accordingly, the degree of policy integration should not be notably different.

The institutional approach would provide a more complicated picture. Corporatism's institutional capacities for intervention and its ability to steer a society may allow it to incorporate environmental interests within decision-making circles. Its capacity for compromise and positive-sum relations would also be assets in integrating environmental interests with economic interests. However, there is also a possibility that a tight policy network of corporatism would be resistant to the intrusion of new environmental interests, thus hampering integration. Pluralism, on the other hand, would be relatively open to environmental interests, but they would need to fight for influence within the winner-takes-all decision-making style. Also, the lack of governmental steering would do little to help any processes of integration.

Stringency

Corporatism's institutional capacity for intervention, for transforming zero-sum relations into positive-sum relations and for concerted action suggest that stringent policy could be taken and implemented effectively. However, the corporatist policy style of compromise may result in a 'second-best' solution. Moreover, corporatism's institutional bias may work against global warming policy. Under pluralism, on the other hand, policy would depend on which group wins the political competition. Even if environmental interests win, however, a lack of interventionist practice and the competitive policy style might undermine the policy over time, in particular at the implementation stage.

Wilson's policy typology indicates that when costs are concentrated on energy-intensive industries they are likely to lead those industries to resist global warming policy, regardless of institutional differences. Also, the issue characteristics of the global warming problem, which constrain governments' ability to tackle it, will make it more difficult for a stringent policy to be adopted. Nevertheless, effective policy catalysts may lead to the introduction of a robust policy.

Conclusions

Although this chapter presents the two approaches in a contrasting way, I do not wish to imply that either of them is correct or incorrect. The two approaches are not mutually exclusive but complementary. Neither do I intend to present the implications of the two approaches as hypotheses to be tested empirically. There are too many possible patterns of explanation to be explored through only two countries. Instead, my aim is to examine in detail which of the possible patterns of relationship appear to be in operation. This will, in turn, help evaluate the two approaches and allow us to consider the relations between the two in the study of environmental policy.

Before proceeding to the next chapter, let me summarise how the concepts introduced in this chapter will be used in the subsequent, empirical chapters (5–8). First, 'corporatism' and 'pluralism' are used as shorthand to denote opposing sets of institutional characters (covering business organisation, government–industry relations and decision-making style). Second, international influences and individuals as policy catalysts are treated as an exogenous stimulus to the domestic policy process. In the subsequent chapters, these concepts will be used in combination to explain and contrast Japanese and British policies on global warming.

Notes

1 Their horizontal and vertical dimensions are in line with Jänicke's concepts of 'intra-policy coordination' and 'inter-policy coordination', respectively (Jänicke, 1997: n4).
2 Hall (1993) and Sabatier and Jenkins-Smith (1993) divide belief systems into three parts. Hall says that policy change can be first-, second- or third-order change, with the last being more significant change, and Sabatier and Jenkins-Smith propose core, near-core and secondary beliefs.
3 A distinction is often made between an authoritarian form of corporatism and societal or neo-corporatism. For simplicity, I will use the term 'corporatism', but it means societal corporatism.

4 They admit variations in styles from sector to sector. Also, although the notion of 'national style' indicates uniqueness, it does not necessarily exclude generality. As Richardson *et al.* themselves propose four types of national style, the notion still allows certain general features and provides opportunities to make a typology of national styles.

5 In this study I will use 'policy networks' as a generic term for various types of policy network as described by various scholars.

6 McEachern uses the Australian experience in arguing this. Given that Australia is basically pluralist, his argument is not free from doubt. Nevertheless, it is worth noting.

7 The EU, in which member states pool their sovereignty on certain issues, is an exceptional case.

4
Making global warming policy

This chapter gives background information on policy-making to tackle the global warming problem in Japan and Britain. In Chapter 3, I briefly explained that Japan could be considered corporatist and Britain pluralist in terms of government–industry relations, patterns of interest representation, and the norm of decision-making. I will elaborate how these differences are actually reflected in the traditionally dominant environmental policy styles of the two countries. Those industrial structural contexts that have important implications for the politics of global warming and the main actors in policy-making over global warming will also be briefly explained.

Policy styles and environmental politics in Japan

Consensus, concertation and developmentalism
Japan underwent a major administrative reorganisation in 2001. Because this study looks largely at politics and policy before then, I will outline the previous organisation. Table 4.1 shows the main ministries and agencies involved in environmental policy. The primary responsibility for environmental policy in Japan lay with the Environment Agency (EA). Its main duty was to coordinate various administrative measures; for the most part it lacked authority for implementation. A number of other ministries assumed jurisdictional competence on sectoral environmental protection and implemented the policies and measures planned and coordinated by the EA (Gresser *et al.*, 1981).

Japanese policy processes are, in principle, based on consensus. Haley argues that the searching for consensus is the effect of the

Table 4.1 *Ministerial responsibilities for the environment in Japan, before 2001*

Ministries and agencies	Environmental responsibilities
Environment Agency (EA) (now the Ministry of the Environment)	Environment problems in general except for energy issues
Ministry of International Trade and Industry (MITI) (now the Ministry of Economy, Trade and Industry) and its Agency of Natural Resources and Energy (ANRE)	Energy issues, industrial pollution and the development of environmental technology
Ministry of Health and Welfare (now the Ministry of Health and Labour Welfare)	Waste disposal
Ministry of Construction (now the Ministry of Land, Infrastructure and Transport)	Urban development planning, implementation of public works such as sewage, city parks, roads and rivers
Ministry of Transport (now the Ministry of Land, Infrastructure and Transport)	Enforcement of standards on automobiles, ships, aircraft and railways
Ministry of Agriculture, Forestry and Fisheries	Management of national forests, pesticides and protection of wildlife
Agency of Science and Technology (now integrated into the Ministry of Education, Culture, Sports, Science and Technology)	The control of radioactive waste based on the Atomic Energy Basic Law. The development of environmental technology
Ministry of Foreign Affairs	Diplomacy and international coordination over global environmental problems
Ministry of Finance	Tax, budgeting and finance (in 1998 financial responsibilities were transferred to the newly established Financial Service Agency)

laws which established Japanese ministries. These laws typically give ministries broad mandates and authority to intervene in and supervise an area of socio-economic life, but with limited powers of enforcement. The weak administrative power of bureaucrats has compelled them to look for consent and voluntary compliance (Haley, 1995; Hall, 1995). Consensus-building between the government and industry is underpinned by a series of formal and informal networks between the two, which itself is well organised horizontally and vertically (Boyd, 1987: 64; Wakiyama, 1987; Stockwin *et al.*, 1988). A relatively well integrated business sector and the extensive business network have not only offered the government points of access, often functioning as the instruments for policy concertation (Hall, 1995), but also contributed to lower barriers of collective action for business in relation to environmental concerns (Wallace, 1995: 107–8; Elder, 2002). To these corporatist characteristics of consensus and policy concertation should be added the legitimacy granted to government in steering the market, which the country's late industrialisation brought about. Developmentalism has institutional expressions in the form of various financial organisations promoting industrial investment, notably the Development Bank of Japan.

These corporatist developmental institutions have been extensively used in environmental policy processes. Engineers have played an important role here, injecting their professional ideas (Gresser *et al.*, 1981). The resultant technological approach attained some success. In 1977, the Organisation for Economic Cooperation and Development (OECD) expressed appreciation of Japanese achievements:

> The Japanese experience in the field of pollution abatement lends support to the idea that to a large extent it is not technology that should constrain policy choice, but policy choices that should constrain technology. (OECD, 1977)

Moreover, Japan has learnt that with a technological approach there are positive-sum relations between the economy and the environment. The world's most stringent emission standards for cars enhanced the competitive edge of Japanese car manufacturers. Japan also leads the world in technologies for the control of sulphur dioxide.

Pitfalls in environmental policy

The gap between the practice of sectorisation of policy-making in liberal democracies and the holistic nature of the environment inevitably imposes constraints on coherent environmental policy. Environmental policy entails interference with and policing of the business of other ministries. This causes jurisdictional turf disputes, frustrating the work of environmental administration (McCormick, 1991: 14). In Japan, the fact that authority to enforce environmental regulations often resides in other government ministries frustrates the EA. Moreover, some of the political characteristics of Japan not only make sectorisation very rigid, thereby undermining environmental policy, but also pose further constraints on environmental policy-making.

First, Japanese politics is characterised by strong ministerialism, which Muramatsu (1994: 25) argues is the core of pluralistic conflict in Japan (whether or not it is inefficient or healthy conflict) (see also Schwartz, 1998: 19–20). In Japan nearly all sectors of the country's socio-economic life and interest groups are divided into the 'natural constituencies' of particular ministries and agencies (Pempel and Muramatsu, 1995: 53; Okimoto, 1989). The tendency for a single-ministry affiliation by interest groups has reinforced sectorisation of policy processes and often hampered the formation of flexible coalitions (Okimoto, 1989: 202; Muramatsu and Krauss, 1987). Moreover, political leadership is typically weak. The setting of specific policy goals and priorities is largely left to bureaucrats. This has made Japanese policy processes more decentralised than otherwise.

The second characteristic and related to the weak political leadership is the way in which the ruling LDP has been involved in policy processes. The LDP has exerted its influence by responding to more particularistic demands from interest groups (Okimoto, 1989: 210). The main mechanism for this function is the *zoku*, or 'policy tribes', which divide the LDP horizontally, in addition to the vertical divide produced by a factionalism based on patron–client relations between politicians. A *zoku* is a loose and informal group of influential Liberal Democrats who have a special interest in a specific policy area. *Zoku* have been instrumental in mediating among various interest groups and the bureaucracy, and contributed to the creation of a series of policy networks centring on ministries. What is important here is that these policy networks have developed interlocking self-interests among bureaucracy, politicians and interest groups based on reciprocal political exchange. The most crude picture of this political exchange is the 'money politics' model of Japanese politics. Electoral candidates look to business for donations to finance their campaigns, which are based on particularistic goods ('pork-barrel' projects and personal gifts); after election, the candidates reciprocate by pushing bureaucrats to favour business.[1]

The implications of these characteristics of Japanese politics for environmental policy are significant. First, strong ministerialism and tight policy networks indicate that the EA would have difficulty in crossing over to other economic policy networks where the solution to the problem lies, and in influencing their policy outcomes. Second, tight policy networks indicate exclusion of certain groups, and what is notable in Japanese politics is the 'patterned' nature of such exclusion. During the thirty-eight years of reign by the LDP (1955–93), those who did not ally themselves with the party or were not compatible with its policy line – a pillar of which includes the primacy of the economy – were systematically excluded from the centre of policy-making (Muramatsu and Krauss, 1987). Labour and environmental groups were examples (Pempel and Tsunekawa, 1979). These interests have not only been excluded but left disorganised and underdeveloped. And this has further implications for environmental politics. According to Okimoto's perspective on state power in Japan, the EA is severely disadvantaged *vis-à-vis* the economic ministries. Okimoto argues that state power in Japan

> hinges on its capacity to work effectively with the private sector.... Bureaucratic power is ... relational in the sense that it emerged from the structure of LDP–bureaucracy–interest group alignments and political exchanges that take place among them. (Okimoto, 1989: 314)

The EA lacks both political and social relationships. The constraint on environmental policy is described by Broadbent (1996) as 'communitarian elite corporatism'. Environmental policy, especially in the early 1970s, was dominated by the traditional ruling triad (big business, pro-growth politicians and economic bureaucrats), who collectively and pragmatically responded to the public environmental demand as it began to threaten their legitimacy. It is a common feature of liberal democracies that environmental ministries are weak *vis-à-vis* their economic counterparts. In Japan, such power relations are institutionalised.

Policy styles and environmental politics in Britain

Consultation, science and reactivism

After 1997 the incoming Labour government reorganised environmental administration. Table 4.2 shows the main departments involved in environmental policy processes in Britain under the previous, Conservative administration. Administrative responsibility in this area was principally with the Department of the Environment (DOE). The table shows the ministries as they were configured before the Department of Transport merged with the Department of the Environment under the Labour government which came into office in May 1997.

As in Japan, policy processes in Britain are highly departmental and dominated by a number of decentralised policy networks (Jordan and Richardson, 1982). However, unlike in Japan, where the status of 'insider' or 'outsider' groups is fundamentally influenced by their congruity with the policy line of the ruling party (the LDP), in Britain it depends a great deal on groups' capacity and resources, as well as their preparedness to comply with the 'rules of the game' set out by the state.

Policy-making is characterised by extensive consultation with affected groups. According to Kavanagh (1990: 147), it is a 'cultural norm': 'A group expects to be consulted, almost as a right and certainly as a courtesy, about the details of any forthcoming government legislation and administrative change that is likely to affect it'. The aim of consultation is to build consensus. Consultation, consensus and self-regulation have been the traditionally favoured approach, rather than coercion. This is also the case with pollution control.

The British practice of consensus and voluntary compliance is, however, essentially different from the 'consensus' of Japan. Perhaps because the country was the first to undergo industrialisation, British political culture is rooted in a strong liberal ideology (Marquand, 1988). There is no tradition of, or even legitimacy for, public intervention in the market without good reason. To lead and steer the economy is not an appropriate role of the state.

Table 4.2 *Ministerial responsibilities for the environment in Britain, before 1997*

Departments	Environmental responsibilities
Department of the Environment (now Department for Environment, Food and Rural Affairs, or DEFRA)	Environmental protection, including energy efficiency (after the abolition of the Department of Energy in 1992) and local government matters
Ministry of Agriculture, Food and Fisheries (now integrated into DEFRA)	The prevention of pollution from farming activities
Department of Trade and Industry (DTI)	Energy policy, except for energy efficiency, after the Department of Energy was abolished in 1992
Department of Transport (DOT)	The control of marine pollution. The impact of vehicle emissions on air and human health
The Treasury	The introduction of economic instruments for environmental purposes
Foreign and Commonwealth Office	International environmental policy. Global environmental assistance
Department of Health	Health aspects of environment and food

Source: *Civil Service Yearbook* (Her Majesty's Government, 1997a).

Thus, even though consultation in search of consensus-building does take place, as Katzenstein observes:

> The line between state and society is clear. The sharing of information, consultation, and bargaining distinctive of Britain's 'collective politics' takes place at a table which separates public from private sectors. (Katzenstein, 1978: 312)

This contrasts with Japan, where extensive networks between government and industry blur the line between the public and private spheres.

Moreover, unlike in Japan, the private sector in Britain is basically pluralist. Industrial groups are generally strong and there are peak associations; however, these organisations are not well integrated. The main industrial peak association, the Confederation of British Industry (CBI), has rather limited representation and authority over its members. There is also a rival peak association, in the form of the Institute of Directors, and the presence of two peak associations has further fragmented the business sphere. The British organisation of business does not have an institutional affinity to a strategic corporatist policy.

Under such a liberal tradition, 'consensus' is more passive than the corporatist equivalent. There is no element of active support for a policy but rather 'imposition of will by consent' (O'Riordan and Rowbotham, 1996: 228). As there is no state leadership in shaping the economy, this 'consensus' does not lead to proactive policy concertation orchestrated by the state.[2] Indeed, combined with a predilection for 'avoidance of radical policy change and a strong desire to avoid actions which might well challenge well entrenched interests' (Richardson, 1982: ix), the British consultation process has resulted in a reactive (as opposed to proactive) policy style. In the environmental policy area, this has been exacerbated by the 'institutional bias' against engineering in favour of scientists and a commitment to the principle of 'sound science' in making policy (Wynne and Simmons, 2001). Scientists are inclined to conduct more research and establish evidence rather than devise practical solutions, and their advice is based on the belief that 'unproven' things are not a suitable basis for action (Boehmer-Christiansen and Skea, 1991; Weale, 1992: 80–3; Hajer, 1995). The government considers that science should justify intervention, but non-interventionism has been reinforced by science.

Although the British policy style is characterised by consultation, consensus and reactivism, party politics based on majoritarianism indicates otherwise. Political leadership in Britain is stronger than in Japan and major political parties are more cohesive than their Japanese counterparts, especially the LDP. Party policies which directly or indirectly affect the environment have, therefore, significant effects on national environmental policy. National environmental policy can be pursued decisively as a party policy, or marginalised, especially when competing policy objectives are pursued. During the most of the Thatcher years, the Conservative Party marginalised environmental interests and politics.

Industrial structural contexts of global warming politics

In the context of the politics of global warming, the role in the country's economy of fossil fuels (coal in particular – see Chapter 2) is of prime importance; this may be in both fuel-extraction industries and energy-intensive industries. In the absence of a practicable technological solution to remove carbon dioxide emissions, the most urgent policy measures are energy efficiency and switching from fossil fuels, especially coal – the most carbon-intensive products – to alternative sources of energy. Less carbon-intensive fuels should also be promoted, to replace coal and oil. Thus, in the long term, the coal industry (and perhaps the oil industry),[3] and more immediately the energy-intensive sectors, will be most affected by a policy of reducing emissions of carbon dioxide.

Japan and Britain have different energy resources. Japan is a resource-poor country, and so imports more than 95 per cent of its energy supply (ANRE, 2003: 6). About 97 per cent of the domestic demand for coal is met by imports. Japan is the world's biggest coal importer, accounting for a third of the world coal trade (Council for the Coal Industry, 1999: 5).

Britain, in contrast, has abundant supplies of coal, oil and gas. At present it is able to meet over 70 per cent of its requirements for coal and Britain has been a net exporter of oil and gas (these account for about 9 per cent of the country's export earnings).[4] The importance of the oil, gas and coal industries to the economy has, however, rapidly reduced since the 1980s. Recently they have contributed roughly 2 per cent to gross domestic product (GDP) and industrial employment. Oil and gas production is set to decline rapidly over the next decade, and Britain is expected then to revert to being a net importer of oil and gas (OECD and IEA, 2002: 22–6, 81). During the 1990s, the long-term decline of the coal industry was accelerated by privatisation of the electricity industry, the biggest consumer of coal products, followed by privatisation of the coal industry itself. Employment in the coal industry currently stands at about 12,000, down from 60,000 in 1990 (OECD and IEA, 2002: 76). This figure is, however, still much bigger than that in Japan. Under the Japanese government's policy of phasing out the domestic coal industry, employment in coal mining declined from about 7,900 in 1990 to just 777 in 2002 (Statistical Research and Training Institute, 2005: 283). An

Figure 4.1 *Economic structures (value added) in Japan and the UK, 1990 and 2000.*

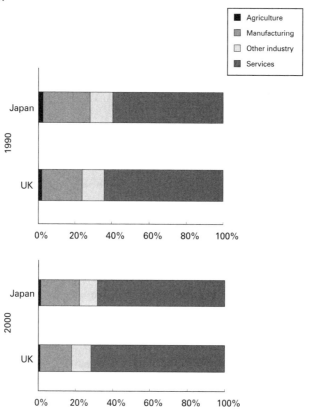

Source: OECD (2001).

important implication here is that Japan has a macroeconomic as well as a political advantage in pursuing global warming policy, as it will help reduce the country's trade deficits, and one of the biggest losers of the policy is virtually absent.

The manufacturing sector in both countries has a diminishing role in the economy. Figures 4.1 and 4.2 show that the manufacturing and 'other industry' sectors play a somewhat larger role in the Japanese economy than the British. As Tables 4.3 and 4.4 show, the role of the energy-intensive sector within the manufacturing sector is not so dissimilar between the two countries, except

Figure 4.2 *Employment structures in Japan and the UK, 1990 and 2000.*

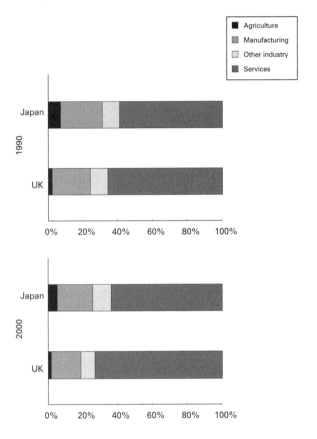

Source: OECD (2001).

for the iron and steel and chemicals industries. Japanese steel makers, though declining, remain some of the most important players in the world market. In Britain chemicals are the largest manufacturing sector and are major export goods.

All in all, neither Japan nor Britain is heavily dependent economically on the industries most affected by climate policy, and these industries are in any case generally in decline. Nevertheless, they do have an important role to play in each national economy, particularly the energy industries in Britain, although in general

Table 4.3 *Output of energy-intensive industries as a percentage of total manufacturing output in Japan and Britain, 1990 and 2000*

	Iron and steel		Non-metallic minerals		Paper and paper products		Chemicals[a]		Non-ferrous metals		Total	
	1990	2000	1990	2000	1990	2000	1990	2000	1990	2000	1990	2000
Japan	8.3	5.9	3.1	2.6	3.0	2.9	6.2	6.4	2.3	1.7	23.0	20.0
Britain	3.8	2.2	3.1	2.8	3.2	2.8	7.7	7.4	2.0	1.3	19.9	16.5

[a] Pharmaceuticals are excluded.
Source: OECD (various years), *The OECD STAN Database for Industrial Analysis.*

Table 4.4 *Employment within energy-intensive industries as a percentage of total manufacturing employment in Japan and Britain, 1990 and 2000*

	Iron and steel		Non-metallic minerals		Paper and paper products		Chemicals[a]		Non-ferrous metals		Total	
	1990	2000	1990	2000	1990	2000	1990	2000	1990	2000	1990	2000
Japan	2.8	2.6	4.1	4.0	2.5	2.5	2.4	2.8	1.4	1.3	13.2	13.2
Britain	2.6	1.9	4.1	3.7	2.3	2.5	4.5	4.5	1.0	1.1	14.7	13.7

[a] Pharmaceuticals are excluded.

Source: OECD (various years), *The OECD STAN Database for Industrial Analysis.*

Figure 4.3 *Energy consumption in industry in Japan, 1990 and 2000.*

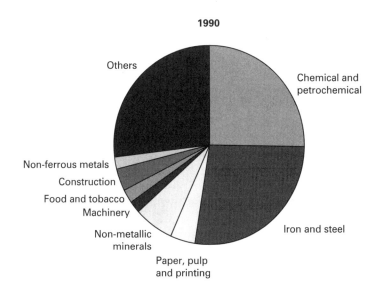

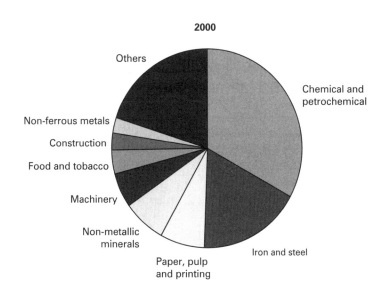

Source: IEA (1993, 2002).

Figure 4.4 *Energy consumption in industry in Britain, 1990 and 2000.*

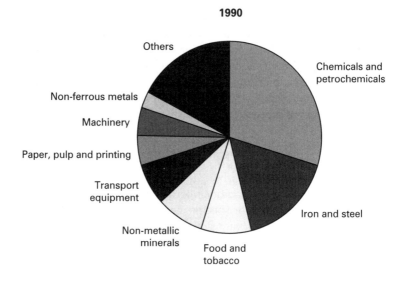

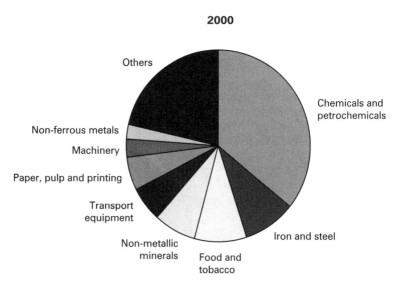

Source: IEA (1993, 2002).

energy-intensive industries in Japan have a bigger economic role than they do in Britain.

As Figures 4.3 and 4.4 show, in both Japan and Britain energy consumption in the industrial sector is concentrated in a few industries. In Japan, the iron and steel and chemical industries are by far the biggest energy consumers – they alone account for more than half of industrial energy use. In British also, the chemical and iron and steel industries represented a substantial proportion of energy use in 1990, but with iron and steel declining.

Main actors in global warming policy-making

Given that the sources of carbon dioxide lie in almost all socio-economic sectors, a large number of actors have an interest in certain aspects of the issue or specific measures proposed. This section does not cover all relevant actors, only the main actors who are either very concerned about or very committed to the policy of reducing emissions.

Japan

The EA was a special department of the Prime Minister's Office. In May 1988, to be able to respond effectively to looming global environmental problems, the Global Environment Protection Room was established, and in 1990 it was upgraded to the Global Environmental Department. This was the first new department since 1974 and soon became one of the most important departments within the EA (Schreurs, 1995). Its political head, the Director-General for the Environment, sat in the Cabinet as the Minister for the Environment, and after July 1989 he or she also acted as the Minister for Global Environment. The position of the Director-General was not usually taken by a high-profile politician. The term of office was generally very short, particularly between 1988 and 1997. Compared with only five Environment Secretaries in Britain, there were fifteen Directors-General in Japan during this period.

Much more powerful in the area of climate politics was the Ministry of International Trade and Industry (MITI). In particular, its Agency for Natural Resources and Energy (ANRE) was responsible for all energy matters, including energy efficiency, and this enabled MITI to play a crucial role in climate politics. The

Ministry of Foreign Affairs (MOFA) was (and is) also involved in the core circles of policy regarding GHG reduction; however, partly because it has not had a clear stance on environmental issues, the MOFA has been much less assertive than the EA and MITI. The Ministry of Transport (MOT), the Ministry of Construction (MOC) and the Ministry of Finance (MOF) also had a bearing on policy on global warming.

Advisory councils for ministers and directors-general typically serve as a locus for corporatist consensus-building (Schwartz, 1998). The 'consensus' that is sought is not only consent and cooperation from affected parties, but also social consensus. As such, advisory councils are typically composed of various social elements, including influential opinion leaders as well as affected economic groups, professionals and academics. However, the dominant actors in the councils are bureaucrats. Council members are selected by bureaucrats and bureaucrats manage the councils' research, have a voice through ex-bureaucrats as council members and draft final reports. Advisory councils are frequently criticised for being a ministry's legitimising tool for decisions already made by bureaucrats and their close social allies. In this light, their independence is indeed severely constrained (Neary, 2002: 116–17). In the context of policy-making over global warming, the EA's Central Environment Council played a central role. However, MITI's Industrial Structure Council and especially its Advisory Committee for Energy (ACE) have had a decisive influence on Japanese policy. It is said that while the EA's advisory bodies have a more independent role to play, MITI's advisory bodies are dominated by bureaucrats.

Climate science in Japan is rapidly growing but has been weak, especially when the global warming problem loomed large. The government had to borrow scientific information from other countries and the lack of original scientific research sometimes resulted in the government becoming reactive rather than proactive (Yonemoto, 1994; Schreurs, 1995, 2001). Nevertheless, research has been undertaken by the MOT's Meteorological Agency and a study group set up by the EA. Policy-oriented scientific and technical research was also conducted by the EA's National Institute for Environmental Studies (NIES), which periodically provided data on which the EA based its own proposals and targets for carbon dioxide emissions.

Environmental groups in Japan have long been marginalised from the main locus of policy-making. This is another weak point in Japanese climate politics. In contrast to the well developed British environmental groups, Japanese environmental groups are small, lack resources and have little political influence. Despite the rapid growth and increasing understanding and recognition of the importance of their role during the 1990s, institutional, political and social constraints still remain (Eccleston, 1995: 151). Thus, compared with the biggest British groups involved in climate politics, such as the Royal Society for the Protection of Birds (RSPB), which has a membership of about 1 million, the World Wide Fund for Nature (WWF), which has about 330,000, and 91,000 in Friends of the Earth (FOE), in Japan even the biggest group in the climate field, WWF-Japan, has well under 50,000 members. Japanese environmental groups also lack resources. WWF-Japan recently had a budget of about £5 million (WWF-Japan, 2003), compared with some £34 million for WWF-UK (WWF-UK, 2003), about £8.7 million for FOE (FOE, 2003) and £57.8 million for the RSPB (RSPB, 2003).

Nevertheless, Japanese environmental groups are rapidly gaining credibility and popularity, as well as policy-makers' attention. Currently perhaps the most important group active in climate politics is the Kiko Network (formerly Kiko Forum). WWF-Japan, the Citizens' Alliance for Saving the Atmosphere and the Earth (CASA), Greenpeace Japan, the People's Forum 2001 and FOE have also been active in Japanese climate politics. The People's Forum 2001, however, became defunct in 2001 after a seven-year operation, largely as a result of financial difficulties.

Business in Japan is relatively well integrated and politically very influential, in contrast to the underdeveloped, long-marginalised environmental groups. In 2002 two of the four peak business associations, Keidanren (Japan Federation of Economic Organisations) and Nikkeiren (Japan Federation of Employers' Associations), were merged to become Nippon Keidanren (Japan Business Federation).[5] Keidanren was (and Nippon Keidanren is) the more politically influential. It had close ties with the LDP and ministries, and represented virtually all the major corporations and more than 100 industry associations. Keidanren's effectiveness in representing industry with one voice in specific matters has been increasingly questioned, but it still acts as an authoritative business representative and plays a quasi-public role. Among the active industrial associations in global warming politics, the Federation of Electric Power Companies – the industry which represents about a quarter of the total carbon dioxide emissions by source[6] – the Japan Iron and Steel Federation, and the Japan Automobile Manufacturers' Association have a high profile in Keidanren and one or other has effectively held the chair of Keidanren since 1980. The Petroleum Association of Japan has a natural concern with government climate policy, but the industry is relatively fragmented and does not enjoy the high status within Keidanren of the three aforementioned associations. Major industrial associations in Keidanren generally act as the authoritative representative of the industry they represent. They also individually have very close ties with MITI or other economic ministries that have jurisdiction over them. According to the subcommittee of the Central Environment Council, about 60 per cent of total industrial carbon dioxide emissions (at the point of use) were accounted for by four energy-intensive industries in 1998: 34 per cent by iron and steel, 12 per cent by chemicals, 9 per cent by cement and 6 per cent by pulp and paper.

Britain
Under the Conservative government, the DOE had primary responsibility for the environment, but its administrative tasks also included housing and local government matters, which, since its establishment, had been the largest tasks of the DOE and its political head, the Secretary of the State for the Environment, who sat in the Cabinet. Unlike in Japan, over the years, the position of the Environment Secretary was occupied by high-profile politicians, although, given their wide responsibility, this does not necessarily mean that their largest effort was spent on environmental protection. After the abolition of the Department of Energy (DEn) in 1992, the DOE was also responsible for energy efficiency. Other energy matters went to the Department of Trade and Industry (DTI).

One important feature of the British administration is its clear pecking order. Even if high-profile politicians occupied the position of the Secretary for the Environment, the DOE was overpowered by the Treasury and DTI (Weale, 1997: 94). The DOE also faced the political reality that it had to operate in the broader political framework of the government's preoccupation with the economy.

Advisory bodies in Britain are more independent and concentrate more on policy scrutiny and advice than is the case in Japan, where many advisory councils belong to a specific ministry or agency (see above). The Royal Commission on Environmental Pollution (RCEP) is the most authoritative body for environmental policy and administration. Parliamentary committees have also had important influences on government policy. During the 1990s the government established a number of new bodies, including the Advisory Committee on Business and the Environment, which was jointly sponsored by the DOE and the DTI, the British Government Panel on Sustainable Development, which reported directly to the Prime Minister, and the Round Table on Sustainable Development, which brought together a wide range of views and interests, including from environmental groups.[7]

In addition to these rich intellectual sources for policy formulation, Britain also enjoys an internationally acknowledged excellence of its *climate science*. The Tyndall Centre at the University of East Anglia, a world-class centre for climate research since the 1970s, and the newly established Hadley Centre at the UK Met Office, which, until April 2002, housed the Technical Support Unit for IPCC WG-I, have given the best available information on climate change, over the world as well as in Britain. John Houghton, under whose co-chairmanship IPCC WG-I finalised three assessment reports, was from the UK Met Office. Since April 2002 the Hadley Centre has housed the Technical Support Unit for IPCC WG-II and its co-chair. The DOE also had its own scientific review group, the Climate Change Impact Review Group. It is noteworthy, however, that in contrast to Japan, where sceptical scientific voices have been heard little since the early 1990s, in Britain strong sceptics regarding climate change remain and these scientists have been active in voicing their views.

Environmental groups in Britain are stronger than those in Japan (Lowe and Goyder, 1983; Boehmer-Christiansen and Skea, 1991: 80–1; McCormick, 1991: 34). British environmental groups have expertise, a capacity for drafting bills, and highly developed skills in lobbying. FOE, for example, made the first significant and systematic assessment of different response options for carbon dioxide abatement and this remained a major reference point for several years (Wynne and Simmons, 2001: 104). In the early 1990s, however, it was still said that access to the critical economic departments, such as the Treasury, the Ministry of Agriculture, Fisheries and Food, the Department of Transport (DOT) or the DTI remained poor, and despite the steady growth of environmental groups, in particular during the 1990s, these groups were not able to translate their numerical and seemingly increasing power into commensurate real political influence (Ward and Samways, 1992: 122–3). Environmental groups engaged in the climate issue include, *inter alia*, FOE, Greenpeace UK, WWF-UK, the Green Alliance, the Campaign to Protect Rural England and the RSPB. They are broadly accepted as 'established' groups.

Unlike in Japan, there is an established Green Party in Britain. The party has some elected representatives on local councils, but the British fast-past-the-post parliamentary electoral system has penalised small parties like the Green Party.[8] Nevertheless, it has had an important influence on British politics by, *inter alia*, alerting other major parties to environmental votes through its periodical success, especially at local and European levels.

Business in Britain is more decentralised than its Japanese counterpart. The most influential peak trade association, the CBI, has much less integrative capacity than Keidanren. It does not include the financial sector. Large companies such as BP and Shell have by themselves a strong influence on the government. Some industrial associations are well organised, however, such as the Society of Motor Manufacturers and Traders and the Chemical Industries Association. The role of mineworkers is also worth noting here. With its strong ties with the Labour Party, the National Union of Mineworkers was once very powerful. While it declined considerably over the 1990s, its ties with the Labour Party remain. Some business organisations, such as the Association for the Conservation of Energy (ACE) and the Combined Heat and Power Association, act closely with environmental groups. The nuclear industry also advances the non-carbon attributes of nuclear power. A quarter of Britain's total emissions of carbon dioxide come from power stations.

Effects of the EU on British environmental policy and politics
British environmental policy and politics cannot be understood without placing them in the EU context. The EU is 'the single most

important driver' of British environmental policy developments. Today, most British policy is closely associated with EU policy (Jordan, 2002: xi, xv). The EU has affected the national agenda, enhanced opportunities for environmental groups and introduced higher standards. It altered the fundamental policy paradigm, the general style of environmental policy and policy instruments, giving rise to more preventive, transparent and legalistic policy (Haigh and Lanigan, 1995; Weale, 1997; Lowe and Ward, 1998; Jordan, 2002). The increasing interaction between Britain and the EU also had a considerable impact on the DOE. For example, while its autonomy was reduced as a result of bourgeoning EU environmental policy, its political power *vis-à-vis* other departments increased. More importantly, the EU unexpectedly helped boost the importance of environmental issues within the DOE, which had long poured most of its resources into housing and local government affairs (Jordan, 2002).

While Britain is now deeply Europeanised, it has traditionally been sceptical about EU action on the environment. The consensus among policy-makers has been that international action should be reserved for truly international problems (see Jordan, 2002: 30–1). During the 1970s and 1980s, Britain's emphasis on safeguarding sovereignty, frequent exercise of the veto (see below) and dilution of EU legislation in its implementation were among the factors that gave rise to Britain being called the 'dirty man of Europe'. During the 1990s, however, British attitudes towards the EU became much more constructive and positive.

Since the end of the 1980s, the EU has adopted an increasingly progressive environmental policy. The most important boosts to this were the 1987 Single European Act (SEA), which legalised the environmental competence of the EU,[9] followed by the 1993 Maastricht Treaty, the 1999 Amsterdam Treaty and the 2003 Nice Treaty, which further enhanced competence and opportunities for environmental action. In climate politics the EU acts as an independent player at the international level, and has ratified the FCCC and the Kyoto Protocol, which collectively committed EU member states to a GHG reduction target.[10] However, EU action has often been constrained by some of its institutional provisions.

First, although qualified majority voting (QMV) in the Council of Ministers, which was introduced by the SEA, became the standard procedure for environmental legislation under the terms of the Maastricht Treaty, several exceptions were made, leaving these issues requiring unanimity in voting, which thereby gave each member state a veto over any proposal. Crucially, in terms of climate policy, the exemptions included those issues which are primarily fiscal and those with significant implications for energy choice and energy supply. These exceptions doomed the Council's proposals for an EU carbon/energy tax, and the application of QMV for the decision to ratify the FCCC was disputed by Britain on the grounds of the Convention's implications for energy choice (Haigh, 1996: 177–8).

Another major constraint is the principle of subsidiarity, which was introduced by the SEA. According to this principle, EU action is justified only when individual member states are incapable of acting alone. To put it another way, EU policy is not necessary when measures can be adequately taken at the national level. This principle potentially restricts EU action, especially under the FCCC, which member states individually (as well as collectively) ratified. For example, the energy efficiency programme (SAVE – Specific Actions for Vigorous Energy Efficiency) proposed by the European Commission, which included specific measures such as EU-wide insulation standards for each climatic zone, was considerably weakened by application of the subsidiarity principle (see Haigh, 1996: 174–6, 178–9).

While these constraints were critical in the development of EU climate policy, the EU still had an impact on national climate policy. For example, the setting of energy efficiency standards and labelling for traded products were precisely the areas where the EU should take action. More generally, the EU had provided a broader policy framework and direction, encouraging and ensuring national action. Developments at the EU necessarily affected the national agenda and policy debate.

Conclusions

I shall now summarise the main points of this chapter. First, both Japan and Britain share constraints on environmental policy; indeed, these are common to most countries. Japan and Britain also share a similar politico-institutional context in which economic concerns tend to be magnified in the policy-making process. In Japan there exist interest-based, institutionalised relations between economic bureaucrats, politicians and big business, which have produced pro-business policies. In Britain policy-makers have long been preoccupied with improving economic competitiveness.

In Japan, however, corporatist institutions – specifically, consensus-based decision-making, an integrated business sector, and interventionist and developmental institutions – enable pro-active, concerted action in response to environmental problems. With the help of engineering expertise, Japan has been successful in its progressive technological approach to certain environmental problems. The British policy style is also characterised by consensus and voluntary compliance. However, without an element of active support British 'consensus' is more 'imposition by consent'. Britain is essentially pluralist in its interest representation and lacks a tradition of (and legitimacy for) active state intervention in the market. Traditional reliance on scientists for environmental policy-making served to reinforce such non-interventionism.

The proactive approach in Japan, however, carries a price. Tight policy networks among bureaucrats, politicians and big business create high barriers to inclusion in environmental policy-making for other parties. Moreover, corporatist exclusion in Japan accompanies systematic underdevelopment of environmental groups, especially in terms of their political power. This contrasts with Britain, where there are large, established environmental groups actively operating in the political arena. Even if British environmental groups are less influential than economic groups, it is not because of their systematic exclusion.

Britain also enjoys more established research in natural science and more independent and authoritative advisory bodies than Japan. Japanese climate science was weak in the early years of climate politics, while the main advisory bodies tend to be in the realm of ministerialism, acting as an instrument for forging 'consensus'. An important implication here concerns the institutional environment for policy learning. Extensive networks between MITI and industry in Japan certainly provide an opportunity for learning; however, in terms of environmental policy Britain appears to have greater intellectual resources to rely on.

Finally, perhaps the most important difference between the Japanese and British environmental policy processes is that Britain is part of Europe. British environmental policy and politics are inextricably linked with developments in the EU. Pressure groups have accordingly extended the horizon for their activities and channels for influence.

Notes

1 The multi-seat electoral system in place up to 1996, which generated intra-party competition and reliance on personal support groups, *koenkai,* for electoral campaigns was combined with campaign regulations which hindered policy-based appeals. Even after the changes in the electoral system, candidates still largely rely on a personal appeal for votes. For 'money politics' in Japan, see Cox and Thies (1998).

2 I am indebted for this argument to Albert Weale.

3 According to the IPCC, while the coal industry is expected to face 'almost inevitable' decline, the impact on the oil industry may be moderated, not least by lack of substitutes for oil in transportation. Modelling studies suggest that oil may be the least badly affected and coal the worst affected, with gas probably somewhere between (IPCC, 2001c: 563).

4 The situation is changing particularly quickly in relation to Britain's reserves of North Sea natural gas, which are fast declining: in 2004 Britain was a net importer of gas.

5 The other two are the Japan Association of Corporate Executives and the Japan Chamber of Commerce and Industry.

6 'By source' means before electricity is distributed to end users.

7 In 2001 the incoming Labour government merged the Government Panel on Sustainable Development and the Round Table on Sustainable Development to create the Sustainable Development Commission.

8 The proportional representation system introduced for elections to the devolved assemblies and the European Parliament during the first term of the Labour government brought greater opportunity for the Greens.

9 Before that, environmental issues had been dealt with only in relation to the internal market under the 1957 Treaty of Rome.

10 To be precise, it is the European Community that ratified the FCCC (see Haigh, 1996: 156) but 'the EU' will be used in this book for simplicity and consistency.

5
Policy developments in Japan on global warming: the politics of conflict and the producer-oriented policy response

Japan contributes only about 5 per cent of the world's total carbon dioxide emissions, but this is the fourth highest in the world, following that of the USA (around 25 per cent), Russia (7 per cent) and China (14 per cent). Although Japanese effort to reduce its emissions can make only a marginal difference, it bears an important responsibility in taking part in the world's effort to tackle global warming.

Carbon dioxide accounts for about 90 per cent of GHG emissions in Japan, and about 90 per cent of these carbon dioxide emissions stem from energy-related sources. The historical growth in these emissions is due largely to the growth in energy demand rather than the pattern of use of fossil fuels. The oil crises in the 1970s, and government's and industry's consequent efforts to contain energy demand and to switch from oil to nuclear power and natural gas (Figure 5.1), contained the growth in carbon dioxide emissions. However, declines in energy prices and high economic growth from the mid-1980s stimulated energy demand; the residential and commercial and transport sectors, in particular, grew rapidly (Figure 5.2). It should be noted that, at this time, Japan was successful in decoupling economic growth and carbon dioxide emissions (Figure 5.3), largely through energy efficiency and the switch from oil. However, thereafter, and especially from the mid-1990s on, growth in energy consumption and carbon dioxide emissions has been consistent with that of GDP. Figure 5.4 shows more recent trends in carbon dioxide emissions.

Figure 5.1 *Japan's primary energy sources (percentages as oil equivalents), 1973–2000.*

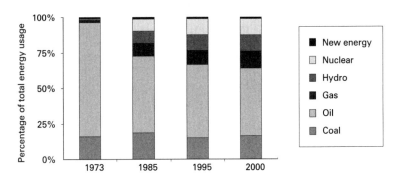

Source: Energy Conservation Centre (2003).

Figure 5.2 *Energy-related carbon dioxide emissions in Japan, 1970–2000.*

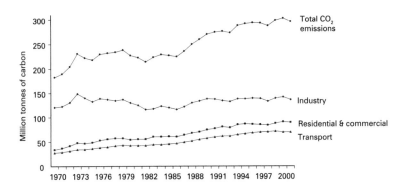

Source: Energy Conservation Centre (2003).

Figure 5.3 *(opposite, top) Trends in GDP, energy consumption and energy-related carbon dioxide emissions in Japan, 1970–2000.*

Figure 5.4 *(opposite, bottom) Japan's total and per capita carbon dioxide emissions, 1990–2003.*

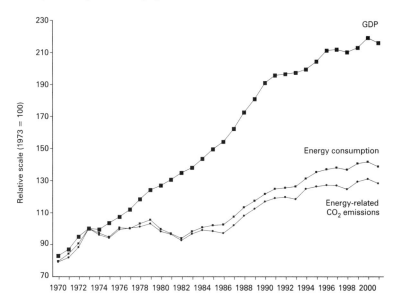

Source: Energy Conservation Centre (2003).

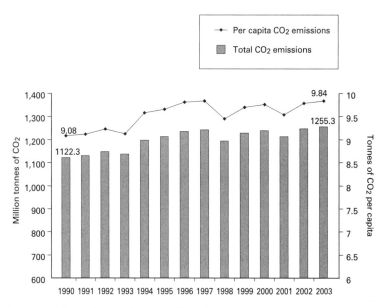

Source: Ministry of the Environment (2004a).

The emergence of global warming on the Japanese political agenda

The global warming problem hit Japanese politics in 1988. The strong governmental interest in the issue thereafter was largely due to the new political climate brought about the abrupt green conversion (superficially at least) of the then Prime Minister, Noboru Takeshita, after the Toronto conference on global warming in June 1988 (see Chapter 2). It is said that he was struck by how environmental issues were being seriously discussed by world leaders. Coming back from Toronto, he enthusiastically and effectively encouraged the LDP to take environmental problems seriously (interview with Saburo Kato, 2000). In 1989 the Japanese government established the Ministerial Council on Global Environmental Protection and the new post of Global Environment Minister. The Council soon formulated the basic policy direction on Japan's role *vis-à-vis* the global environment (see Box 5.1). In 1990 the Council created the Global Environment Research Fund under the EA.

The Japanese government's enthusiasm for global environmental protection should be seen in the context of the broader political debate of the day.[1] The global environment was linked with the problem of Japan's role in international politics. Japanese governments had been domestically and internationally urged to make international contributions commensurate with Japan's economic power. The environment was thought to be a suitable field for Japanese contributions because it is outside the realm of military affairs (in which Japan's action is constitutionally restricted)[2] and because Japan had been successful in the past in overcoming serious problems of pollution with a technological approach. Moreover, as the six-point policy statement indicates (see Box 5.1), part of the solution to global environmental problems was seen as the dissemination of Japanese environmental technology, which thus linked the environment with a new business opportunity. This political context and the issue-framing helped Takeshita's appeal to resonate in domestic politics.

The LDP's environmental zeal reached its peak in 1992. The party created the Special Committee on Global Environment in 1989, and in February 1992 the Investigatory Committee on Basic Environmental Problems, which brought together high-profile Liberal Democrats. This is widely interpreted as the emergence

Box 5.1 *A six-point policy statement from the Ministerial Council on Global Environmental Protection, defining Japan's role* vis-à-vis *the global environment*

The Japanese government undertakes:

1 To use its economic and technological resources to participate in the formation of an international framework for global environmental conservation commensurate with Japan's position in the world community.
2 To promote observation, monitoring, and research.
3 To disseminate technologies contributing to global environmental conservation.
4 To increase Official Development Assistance (ODA) to help developing countries protect their environments.
5 To give greater consideration to the environmental impact of ODA grant and project aid.
6 To work towards establishing economic and social programmes that have lower impacts on the global environment.

Source: EA (1990b: 64–6).

of a new, powerful environment *zoku*. The implications for the EA were considerable. Never before had the EA had such strong political support from the LDP. Moreover, recognition of the economic opportunities afforded by global environmental problems meant that opposition from economic interests was not as fierce as that which met the first wave of environmentalism in the 1970s (interview with Saburo Kato, 2000). The political climate was unprecedentedly favourable for the EA (Schreurs, 1995).

Developments in setting a target

Although there emerged a broad agreement on the need to address the global warming problem within government, there was no consensus on the extent of the actions to be taken. In particular, the policy of target-setting in carbon dioxide control always generated fierce bureaucratic disputes.

EA versus MITI, bureaucrats versus the environment zoku: target towards 2000

Despite Japan's public statement that it was enthusiastic about taking leadership, Japan was undetermined when the world began to discuss the carbon dioxide reduction policy early in 1989. The government could not narrow down the differences in opinions between the EA and MITI, in particular. In the run-up to the Noordwijk conference in November 1989, the clash between the EA, MITI and MOFA sharpened over the Dutch proposal for the Noordwijk Declaration, which included the commitment to the stabilisation of carbon dioxide emissions at the 1990 level by 2000 and a 20 per cent reduction by 2025. While the EA was supportive, MITI and MOFA strongly opposed it. MITI, in particular, emphasised the negative effects on the world economy based on its study. The Japanese economy was considered especially vulnerable since it had already achieved a high level of energy efficiency, hence further improvement would incur greater marginal costs. In the absence of support from the largest producer of carbon dioxide, the USA, any action would not be effective. MITI, together with MOFA, also stressed the importance of good relations with the USA (especially at a time when US–Japanese trade friction was intensifying), which have been the core of Japanese foreign policy. Against these views, the EA argued that it would be possible to achieve both economic growth and carbon dioxide reduction if prudent countermeasures were taken. The EA also emphasised the need to take a leadership role in this field of international politics (*Asahi Shimbun*, 31 October 1989; *Yomiuri Shimbun*, 1 November 1989). Having failed to form a consistent national position on the Noordwijk Declaration, the government adopted a temporary policy: although Japan principally supported a reduction of carbon dioxide emissions, it maintained that the specific target should be discussed after the IPCC had produced its first report (*Nihon Keizai Shimbun*, 4 November 1989).

Japan was severely criticised at Noordwijk for following the USA, but the conference triggered important policy developments in Japan. First, the decision that the Second World Climate Conference, in October 1990, would start negotiations over the creation of the FCCC provided a deadline for formulating a coherent national position. Second, and more importantly, severe criticism of Japan at the conference provoked a number of environmentally enlightened Liberal Democrats to tip the power balance between the EA and MITI in favour of the former. The conclusion of the first scientific assessment by the IPCC in May 1990 (Houghton *et al.*, 1990) provided further support for the EA. Not only did this first assessment report call for immediate political action, but it also brought about changes in the international political landscape (IPCC, 1994). The quick policy change by Britain to have a carbon dioxide limitation target, in particular, isolated Japan, since Britain was the only major developed country, apart from the USA, which had opposed the setting of a target.

At home, MITI's Advisory Committee on Energy, which started to review long-term energy policy at the end of 1989, concluded the Long-Term Energy Supply and Demand Outlook (henceforth Energy Outlook) in May 1990. The report projected that although carbon dioxide emissions would increase by 16 per cent from 1988 to 2000, thereafter they would more or less stabilise until 2010 (*Mainichi Shimbun*, 15 May 1990). With this, MITI, for the first time, agreed to establish a stabilisation target. There remained a gap with the EA, though, which concluded in a report that carbon dioxide stabilisation in 2000 at the 1988 level was possible. Nevertheless, the shift in MITI's attitude enabled the Cabinet to agree to create the Action Programme to Arrest Global Warming with a target of carbon dioxide stabilisation.

Negotiations between the EA and MITI became deadlocked; nevertheless, with general LDP support for the EA, the target was decided just before the Second World Climate Conference in October (interview with Saburo Kato, 2000). To reach the decision, the EA made concessions on changing the base year from 1988 to 1990; the MOT offered to save more carbon dioxide emissions than was estimated in the Energy Outlook; the target was made ambiguous; and two stabilisation targets were set, one on a per capita basis and the other relating to total emissions. They were:

- the stabilisation of carbon dioxide emissions on a per capita basis in 2000 and beyond at about the same level as in 1990;
- the stabilisation of total carbon dioxide emissions in 2000 and beyond at about the same level as in 1990, through progress in the development of innovative technologies, etc., at the pace and on a scale greater than currently predicted (EA, 1990a).

MITI regarded the per capita target as the core and the other as only an aspiration, while the EA emphasised the latter (*Asahi Shimbun Evening*, 23 October 1990). However, because the total reduction was conditional on the rapid development of innovative technologies, the per capita basis was widely seen as Japan's primary target.

The 1990 base year, the per capita basis and the words '*about the same level*' allowed MITI to make virtually no concession from its position based on the Energy Outlook. It should be noted, however, that MITI had, in a sense, already made concessions. Extensive improvements in energy efficiency and an expansion of nuclear power and renewable energy had already been incorporated into the Energy Outlook.

The enactment of the Action Programme was made conditional on similar efforts by the other major industrialised countries; hence the policy change was rather rhetorical, without accompanying real change. This was not peculiar to Japan, however. Many countries which had established carbon dioxide limitation targets made them conditional on similar action by their economic competitors. Nevertheless, the formation of the Action Programme was certainly a watershed in Japanese global warming politics.

Despite the establishment of the carbon dioxide stabilisation target, Japan did not change its opposition to the incorporation of specific targets in the FCCC. It is said that MITI wanted the targets in the Action Programme to lack any legally binding force as far as possible, in case of a failure to meet them (Fermann, 1992: 46). However, the government's proposal of a pledge and review (P&R) system at the second session of the INC (INC-2) in June 1991 triggered developments in the Japanese position. P&R is a system in which each country is obliged to pledge a national strategy to limit carbon dioxide emissions and then non-national specialists will review progress. Similar proposals were also advanced by Britain and others. At INC-2, however, international NGOs strongly criticised P&R and attacked Japan and other proposing countries for not incorporating a numerical target in the Convention and following the negative USA. The Japanese media reported that Japan was being internationally severely denounced (Yonemoto, 1994: 120–1).

This provoked the LDP's environment *zoku*, because bureaucrats had not informed the then Director-General for the Environment or the LDP's Special Committee on Global Environment of its intention to propose P&R (*Asahi Shimbun*, 24 July 1991). Moreover, there was some confusion between the EA, MOFA and MITI about the interpretation of P&R. While the EA and MOFA interpreted it as a tool to contain carbon dioxide emissions (the target to be negotiated), MITI presumed that countries would also voluntarily pledge carbon dioxide stabilisation targets. For MITI, which wanted to bring the USA into the FCCC, this type of P&R might be useful. However, the LDP's environment *zoku* thought that the bureaucrats' thoughtless move had undermined their earlier achievements in global environmental politics and diplomacy.

After INC-2, the LDP's Special Committee on Global Environment contended that Japan should take a leading role in line with the EU position. It very unusually refused to approve the proposal the bureaucrats had already submitted to the international conference, and demanded a re-examination of it, taking into account the Committee's view (*Asahi Shimbun*, 26 July 1991). In August 1991, the LDP's Special Committee approved the new Japanese proposal. The proposal clarified the obligation imposed on all countries to limit carbon dioxide emissions and on industrialised countries to make a maximum effort towards 'the stabilisation of CO_2 emissions about the 1990 level'. P&R was also approved as a strategy (*Asahi Shimbun Evening*, 28 August 1991). Many of these elements were incorporated into the FCCC.

After the Earth Summit
When, in 1994, the international community started discussing the commitment beyond 2000, Japan was experiencing difficulty containing carbon dioxide emissions (see Figure 5.4). Not surprisingly, Japan was opposed to establishing legally binding targets beyond 2000. By this time the LDP's environmental zeal had also waned, in line with a decline in Takeshita's political influence, increased political turbulence and continuing economic recession, and there was little political reproach against the fact that Japan, as one of the JUSCANZ (see Chapter 2), frustrated the EU's attempt to develop and strengthen the FCCC.

The USA's sudden shift in position in support of a legally binding target at COP2 in 1996, however, forced Japan to reconsider its opposition. The USA's approval enabled the COP to formulate the Geneva Ministerial Declaration, which called for a legally binding target. Japan, or more specifically MITI, together with Canada and Australia, secretly attempted to emasculate the 'legally binding' passage in the Geneva Declaration. However, this was quickly revealed and severely criticised by international NGOs, which in turn provoked strong international and domestic pressure for Japan to take a more positive stance.[3] The pressure was reinforced by the fact that at COP2 Japan was formally designated to host COP3, at which the Protocol was to be decided. It is widely known that the real purpose of hosting COP3 was not in fact related to any environmental issue but rather to the Japanese government's ambition to join the UN Security Council as a permanent member. Nevertheless, once it was decided, the responsibility of being the host country put pressure on the Japanese government not to frustrate positive developments supported by the majority. The deciding factor which moved Japan was, however, the leadership exerted by the then EA Director-General, Sukio Iwadare, an environmentally sympathetic Social Democrat. Iwadare took seriously the pressure from national and international environmental groups and the mass media, and his strong political initiative, as the top decision-maker of the Japanese delegates at COP2, compelled MITI and MOFA to agree to the 'legally binding target' of the Geneva Declaration (interview with Yasuko Matsumoto, 2000; see also Takeuchi, 1998: 122–3).

The inter-ministerial negotiation on Japan's proposal for the legally binding target was slow, and it was only two months before COP3 (held in December 1997) that Japan could formally announce its proposal. The main actors in the decision process were the EA, MITI and MOFA. Again, the process featured fierce competition between the EA and MITI. They disagreed not only on the level of stringency but also on the timing of the decision and the announcement of the target proposal. The EA wanted to announce it as soon as possible, in order to show credibility as the host country of COP3. MITI, on the other hand, placed the foremost importance on bringing the USA into an agreement. It therefore argued that Japan should announce its proposal only after the USA had, so that it could take account of that US proposal (Suwa, 1997: 40–1; Takeuchi, 1998: 158–9; Tanabe, 1999: 122–3).

In December 1996 Japan presented an outline of the Japanese proposal. It was composed of two options:

- to contain per capita carbon dioxide emissions below p tonnes a year on average over the five years from the year $2000 + x$; or
- to reduce total carbon dioxide emissions by q per cent on average from the 1990 level over the five years from the year $2000 + x$.

MITI, which wanted consideration to be taken of different countries' existing degrees of energy efficiency, and the EA, which wanted to ensure a reduction in total carbon dioxide emissions, could not reach agreement, as a result of which specific figures for p, q and x were not presented (Suwa, 1997: 31; Takeuchi, 1998: 137).

On his return from the G8 Summit in Denver, USA, at which the issue of global warming was discussed, and from the UN General Assembly Special Session on Environment and Development in June 1997, Prime Minister Ryutaro Hashimoto urged the government to decide on a Japanese target proposal immediately. At the seventh session for the negotiation for a Protocol at the end of July, rather than formally announce the specific figures for p and q, the government informally negotiated them with the USA and several major European countries. This resulted in a diplomatic failure. Not only was this secretive Japanese approach leaked and attacked, but also the proposal itself was too weak to meet the new international norm, 'significant reduction', as incorporated in the Geneva Ministerial Declaration. In fact, Japan allowed itself to increase its total carbon dioxide emissions beyond 2000 (Takeuchi, 1998: 151–5; Tanabe, 1999: 116–21).

In August, Hashimoto again urged the EA, MITI and MOFA to come up with a formal Japanese target proposal, and directed them to decide it before the USA announced its target. MITI's argument was, thus, overruled, and the decision-making process was facilitated. Moreover, Hashimoto implied the need for at least a 5 per cent reduction from the 1990 level in 2010 for the success of COP3. This no doubt gave some support to the EA and put pressure on MITI. The EA, based on an NIES report, indicated that a 7–8 per cent reduction by 2010 was possible with further improvements in energy efficiency, expansion of 'new energy' sources (see below) and the introduction of an environment tax. MOFA, taking the ultimate objective of the FCCC seriously, proposed a 6.8 per cent reduction by 2010. MITI, however, insisted that to contain *increases* in carbon dioxide emissions to 3 per cent was the strictest target possible, based on its own study, which assumed no new big measures.

The different positions of the EA and MITI were especially evident in their projections on the economic damage likely to be inflicted by a 5 per cent reduction in carbon dioxide emissions. According to MITI officials, this would have dire socio-economic consequences, including a 5 per cent reduction in the output of large companies, a ban on the manufacture of electrical goods with a 'stand-by' function and various energy austerity measures, all of which would result in about 1.9 million job losses by 2010. The EA's projected picture was much more optimistic. A 5 per cent or more reduction in emissions was compatible with increased quality of life and economic growth, provided there was dissemination of new technologies and more waste recycling. People's socio-economic life would little change except that cars and electrical goods would have to become more energy efficient (*Asahi Shimbun*, 17 October 1997).

Meanwhile, public interest in Japan's proposal increased. A year before COP3, thirty-six Japanese environmental NGOs got together to establish the Kiko Network, with the aim of pressuring the government towards COP3, and this raised public awareness of the issue. Politicians' interest also grew. GLOBE Japan (Global Legislators' Organisation for a Balanced Environment) held a symposium on global warming policy. Also, environmentalists from the three ruling parties (the LDP, the Social Democratic Party and Sakigake) set up a project team to pressure the government to take a positive stance at COP3. At the same time, industry also made a move. Keidanren asked the government to pursue a flexible and realistic target and opposed the introduction of an environment tax. MITI was, in particular, under strong pressure not to establish a stringent reduction target (*Yomiuri Shimbun*, 9 October 1997; *Enerugî to Kankyô*, 2 October 1997).

On 6 October 1997, the government announced its target proposal. The result of secretive ministerial negotiations was, overall, a 'victory' for MITI. During the negotiations, MITI had to concede its insistence on a 0 per cent reduction in the face of a general concern about the need to live up to the international norm and to meet Japan's responsibility as the host country for COP3. Nevertheless, its detailed data and information about opportunities for energy efficiency and fuel switching prevailed over the EA's arguments (Takeuchi, 1998: 165–6). As a result, a 5 per cent reduction was maintained as a 'base reduction rate', but through differentiation methods based on emissions in proportion to GDP, per capita emissions and population growth, the targeted actual reduction for Japan was 2.5 per cent. The GHGs covered in the proposal were carbon dioxide, nitrous oxide and methane. The incorporation of nitrous oxide and methane enabled small extra reductions to be called for. A still resistant MITI further proposed a flexible interpretation of the target, making the Japanese reduction target only 0.5 per cent, with no reduction for energy-related carbon dioxide emissions. Although the flexible interpretation was not officially adopted, MITI repeatedly insisted that the actual domestic target would be a 0.5 per cent reduction (Takeuchi, 1998: 166–8, 191).

Just after the government's announcement, an environmental NGO, CASA, published a report that showed a 21 per cent reduction in carbon dioxide emissions by 2010 from the 1990s level was well within reach with a robust policy strategy, but without reducing the living standard from that of 1995. It further argued that the benefits of energy efficiency would surpass costs by ¥41 trillion between 2000 and 2010 (*Asahi Shimbun*, 8 October 1997). Other environmental groups such as the Kiko Network, Greenpeace Japan and WWF-Japan strongly objected to the government's proposal. The opposition parties and the two ruling parties, Sakigake and the Social Democratic Party, asked Prime Minister Hashimoto to withdraw the proposal. They objected especially to the secretive nature of the decision-making process. The government's decision was made without any consultation with the ruling parties' project team for COP3, and the government did not even inform environmentalists in the ruling parties, let alone the public, of the Japanese proposal before it consulted with the USA. The ruling party's project team was allowed to re-examine the Japanese proposal, but, after all, could not reach a consensus on a more stringent target, as Sakigake and the Social Democratic Party demanded (Takeuchi, 1998: 170–3).

National strategy

The national strategy to tackle global warming was set forth in the Action Programme to Arrest Global Warming. Action Programme

Box 5.2 *Policies and measures to limit carbon dioxide emissions under the Action Programme*

Formation of urban and regional structures with low carbon dioxide emissions:
• increase of greenery in cities to alleviate the 'heat island' phenomenon
• promotion of energy-saving buildings
• introduction of electricity co-generation systems
• utilisation of unused heat from urban activities, such as subways, through usage of heat pumps
• diffusion of district heating systems
• supply of heat from waste incineration
• utilisation of energy from sewage sludge.

Formation of transport systems with low carbon dioxide emissions:
• reduction of carbon dioxide emissions from individual motor vehicles
• increase in the energy efficiency of trains, ships and planes
• introduction of automobiles with low carbon dioxide emissions, including electric cars
• modal shift to mass transit systems, such as railways and ships, in areas of medium- and long-distance transport between major terminals
• improvement of the transport efficiency of lorries
• construction of bypasses, ring roads and other facilities to mitigate traffic jams
• facilitation of sophisticated traffic control systems.

was a product of secretive ministerial negotiations. There was no open consultation, let alone public debate.

Industry, represented by, for example, the Federation of Electric Power Companies, Keidanren and the Japan Automobile Manufacturers' Association, had expressed its concern about the government's response to the global warming problem before the negotiations over the Action Programme started. Industry stressed Japan's high level of energy efficiency, hence the difficulty

Formation of production structures with low carbon dioxide emissions:
• improvement of combustion efficiency
• introduction of energy-saving manufacturing facilities and production processes
• improvement of the energy efficiency of farming machinery and fishing boats, among others, and the use of natural energy in agriculture, forestry and fisheries
• improvement of the energy efficiency of construction machinery in the construction sector.

Formation of an energy supply structure with low carbon dioxide emissions:
• improvement in the efficiency of power generation
• development and use of nuclear power, with an assurance of safety
• use of hydraulic and geothermal energy, photovoltaic and wind power systems, and natural gas
• introduction of dispersed power generation, such as fuel cells and photovoltaic cells
• development of the infrastructure for the urban use of LNG
• reduction in the differences in demand for electricity between day and night.

Realisation of lifestyles with low carbon dioxide emissions:
• recycling
• a review of excessive packaging
• use of products with low carbon dioxide emissions
• introduction of a 'daylight saving system' or summer time
• reduction of working hours
• appropriate temperature adjustment in air conditioning and heating
• introduction of equipment that is highly energy efficient.

Source: EA (1990a).

in cutting carbon dioxide emissions further, and it was opposed to a uniform carbon dioxide abatement target. Nevertheless, while calling for research to reduce the scientific uncertainty, industry cautiously accepted the need to take feasible measures – from the promotion of nuclear energy and research and development (R&D) of climate technologies, to energy efficiency and, especially, technology transfers. When the Action Programme was decided, industry was doubtful about the attainability of the carbon dioxide stabilisation target but, in general, expressed its intention to cooperate with the government. The Japan Iron and Steel Federation, the Japan Automobile Manufacturers' Association, the Federation of Electric Power Companies and the Japan Gas Association were ready to invest in carbon dioxide reduction measures. The gas industry, in particular, found in global warming policy a new business opportunity to expand markets in liquefied natural gas (LNG). The Petroleum Association of Japan also indicated its intention to work on reducing carbon dioxide emissions; however, pointing to the level of scientific uncertainty, it posed fundamental questions about global warming policy, and this invited severe reproach from the LDP's environment *zoku* (*Enerugî to Kankyô*, 15 November 1990, 6 December 1990, 13 December 1990).

The Action Programme expressed government's intention to work on technological breakthroughs as a way to 'harmonise' environmental and economic objectives. It was essentially a policy menu (Box 5.2), and most policies on it were already being pursued for other reasons. A specific reduction quota for each measure was not allocated, as there was a fear that specific figures might be taken as an international pledge (Suwa, 1997: 5).

The Action Programme was gradually developed and implemented by the competent ministries, especially after the government signed the FCCC at the Rio Summit in 1992. The main policy developments and measures are discussed below.

Energy efficiency policy

In 1993, an energy efficiency policy package was introduced as a central plank of global warming policy. Its main feature was a series of revisions to the Energy Conservation Law, which had been introduced in 1979 as a countermeasure to the oil crises, and the Energy Conservation and Recycling Assistance Law (hereafter Assistance Law), which provides financial assistance to firms engaging in environmental protection – energy efficiency, recycling and reduction of the use of chlorofluorocarbons.

The main revisions to the Energy Conservation Law related to the industrial sector. The more important included: a new obligation on specified factories annually to report their energy use status (under the original 1979 Law factories were required only to record this); and the establishment of an indicative target of a 1 per cent annual average reduction in unit energy consumption for factories (the 1979 Law set out only energy efficiency standards). The Law covered factories that accounted for about 70 per cent of energy use in the industrial sector, which in turn constituted about a half of the total energy demand in Japan. Strengthening policy in this sector was, therefore, a sure way to improve energy efficiency.

More generally, the revisions strengthened energy efficiency standards and compliance measures in factories, housing and buildings, electrical appliances and cars. It should be noted that the standards were not legally binding. Factories and business operators were required to 'make an effort' to achieve them. Whether they were complying was evaluated by the competent ministries, in the light of reports or plans submitted by factories and business operators and energy labels on appliances and cars. Only when the ministries judged that their effort was 'extremely unsatisfactory', or that a 'considerable extent' of improvement was necessary, was administrative guidance issued (first a recommendation and then, after consultation with the relevant council, an order); in extreme cases, a penalty might be imposed.

Administrative guidance held the key to ensuring standards were satisfactorily met. However, the absence of criteria for issuing administrative guidance made its actual application difficult. When MITI launched the revisions to the Law, it intended to enhance the level of compliance by strengthening and strictly applying compliance measures (the original compliance measures provided by the 1979 Law had never been used). However, strong opposition from industries, especially the iron and steel industry, and their sponsors within MITI itself frustrated this effort. Moreover, when the revised law was enacted, Keidanren demanded that MITI respect industry's voluntary action, stressing that too much state intervention would undermine moral and any willingness to innovate.

Instead of relying on compliance measures, MITI used various financial incentives to enhance the effectiveness of the Energy Conservation Law, notably the Assistance Law and measures to encourage the application of energy efficiency technologies. Financial institutions such as the Japan Development Bank and the Government Housing Loan Corporation increased the loans available for energy efficiency projects, and special tax measures applying to investments in energy efficiency were strengthened or expanded. In order to ensure finance for these measures, in 1993 MITI reformed its Energy Special Accounts, which were financed by various earmarked energy taxes, by creating a new fund specifically for energy efficiency. MITI also encouraged industry to take voluntary action. In October 1992 MITI issued administrative guidance to eighty-seven industrial associations to formulate a 'Voluntary Plan Concerning the Environment', which should incorporate a commitment to establish institutions to monitor energy use, targets for energy efficiency and/or carbon dioxide reduction, and policies and measures to achieve them. Within a year companies representing about 60 per cent of total manufacturing sales had created 'Voluntary Plans'.

MITI's energy efficiency policy package was, thus, composed mainly of soft interventionist instruments and various financial incentives to encourage industry to invest in energy efficiency technologies. According to MITI, however, a third of designated factories did not take satisfactory measures and many new houses did not meet insulation standards (ANRE, 1997: 20).

New energy

In the first half of the 1990s two policy frameworks for the promotion of new energy were established (on the definition of new energy, see below). One was the New Sunshine Programme, an R&D programme for innovative energy/environment technologies. This Programme was formulated by MITI in 1993, and under it ¥500 billion was to be invested up to 2020. Another policy framework was the Basic Guidelines for New Energy Introduction, decided by the Cabinet in December 1994. It was the first basic policy indicated by the government for the introduction of new energy. The Guidelines for the first time defined 'new energy', although it was widely considered to lack clarity: new energy is 'forms of energy alternative to oil, that are reaching the commercialisation phase from a technological standpoint, but have not yet become widely used due to economic barriers'. Low-emission vehicles, which run on fuels other than petrol or diesel – specifically, electricity, natural gas, methanol or liquified petroleum gas (LPG), but also so-called 'hybrid' vehicles, which can also use petrol – were also included in the list of the new energy to be promoted. The Guidelines designated fourteen forms

Table 5.1 *The main targets for new energy in the Basic Guidelines for New Energy Introduction*

	1992 (actual figures)	2000	2010
Photovoltaic solar power (10,000 kW)	0.36	40	460
Wind energy (10,000 kW)	0.2	2	15
Energy from waste (10,000 kW)	36	200	400
Energy from co-generation (10,000 kW)	277	1,452	1,912
Low-emission vehicles	2,000 vehicles	490,000 vehicles	2.44 million vehicles

Sources: ANRE (1994b, 1997: 26).

of new energy for the time being and set forth targets and policies for each (Table 5.1).

Since 1990 MITI – specifically its affiliated bodies, the New Energy and Industrial Technology Development Organisation (NEDO) and the New Energy Foundation (NEF) – and other ministries have introduced various financial incentives, including subsidies, tax incentives and low-interest loans, to promote the installation of 'new energy' power facilities.

With these various financial incentives, the source of new energy that benefited most was photovoltaic solar power. Aggressive measures to promote its use started in 1992. MITI, through NEDO, launched a subsidy programme for the installation of photovoltaic power systems in public facilities, in which it covered two-thirds of the installation costs. In 1994 a similar measure was introduced for individual households, through the NEF. The aim was to install the systems on 70,000 roofs by 2000, for which subsidies of 50 per cent would be available. The programme received a good response and its budget has steadily expanded. Between 1992 and 1998 the cost of photovoltaic power generation declined by two-thirds, and installed capacity grew from 19 MW in 1994 to 56.9 MW in 1998 (NEF, 2001a,b). Media reporting, national explanatory meetings and consistent efforts by manufacturers and at sales outlets have also increased public awareness of the technology (Energy Information Administration, 1996). Other ministries also became active in promoting new energy. Among the notable examples was an approach attempted by the EA. The EA concluded a voluntary agreement with manufacturers of vending machines to install photovoltaic power systems on 10 per cent of their machines as a pilot project, though this one-year project had to be terminated as it proved not promising.[4]

In 1997, as a further push towards the establishment of the new energy market, MITI expanded the programme to housing complexes, and increased the budget fivefold in comparison with what it had been in 1994. These aggressive policies stimulated private investment in R&D. It is estimated that, in the fiscal year

Figure 5.5 *Voluntary purchase of surplus electricity generated from new energy, 1992–98.*

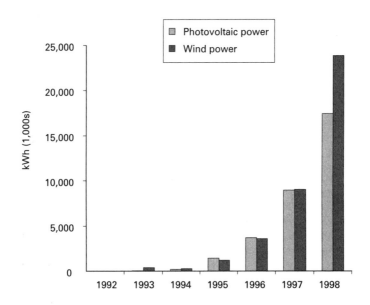

Source: Federation of Electric Power Companies, www.fepc.or.jp/thumbnail/new-energy/surplus.html (accessed 16 July 2005).

1997, the R&D spending of those under MITI's jurisdiction alone exceeded ¥10 billion (Shoda, 1998). With all these developmental measures, Japan achieved the biggest share in the world market for photovoltaic power.

While MITI identified cost barriers as the main constraint on increases in new energy, and tried to reduce costs through mass introduction of new energy technologies using various financial incentives, electric power companies started to purchase surplus new energy power on a voluntary basis in April 1992. Photovoltaic power and wind power have been especially favoured, and purchased at the same electricity price as that charged to industrial consumers. From almost nil in 1992, the amount of photovoltaic power voluntarily purchased increased drastically (see Figure 5.5).

Despite these measures, progress was too slow compared with the ambitious target set for 2000. As a further push, MITI drew up the New Energy Bill and it was quickly passed in April 1997. The Law established a basic policy framework by outlining obligations for governments, energy suppliers, users and new energy producers to strive for the introduction of new energy, and expanded financial assistance measures. While the earlier measures centred on assistance for local governments, the new law offered a new business opportunity for private enterprises. This law will be discussed further in Chapter 6.

Voluntary action

Keidanren has steered the business world in global environmental policy. In April 1991 it issued the Keidanren Global Environment Charter, which laid down a basic philosophy on the (global) environment and guidance for corporate activities, both domestically and overseas. It made clear its view that the environment is the foundation of any business activity, as well as the fact that environmentally considerate business activities can even offer the opportunity for healthy economic development, based on trust from consumers (Keidanren, 1991).

In response to the adoption of the Berlin Mandate at COP1 in 1995, and building on its Global Environment Charter, in 1996 Keidanren adopted the 'Keidanren Appeal on the Environment: declaration on voluntary action of Japanese industry directed at conservation of global environment in the 21st century'. The

Box 5.3 *The Keidanren Appeal*

Three goals for environmental protection:
- reconfirmation of 'environmental ethics' for individuals and organisations to honour
- realisation of 'eco-efficiency', a factor needed to reduce the environmental load through improved technology and economic efficiency.
- tightening of 'voluntary efforts' to cope with environmental issues.

Four urgent issues identified:
- global warming;
- the construction of a recycling-based society;
- restructuring the environmental management system and environmental auditing;
- environmental considerations in evolving overseas projects.

As far as global warming is concerned, the Appeal proposes measures to cope with it as follows:

'Making it a basic policy to review the "throw-away economy", structure a recycle-based society [*sic*] and improve energy efficiency and carbon utilization efficiency, we aim to maintain the world's paramount level of environmental technology. We also aim to improve energy utilization efficiency on a global scale through transfer of appropriate technology to developing countries.'

Keidanren Appeal identified global warming as one of the four urgent issues, and seven concrete measures were suggested (see Box 5.3). Based on the Appeal, industrial associations started to formulate a Keidanren Voluntary Action Plan on the Environment and by June 1997 thirty-six industries with 137 industrial associations, in areas ranging from manufacturing and energy to transportation, finance, construction and foreign trade, had established voluntary plans. The participants together accounted for 80 per cent of the total carbon dioxide emissions from industry. Their carbon dioxide reduction targets would mean that emissions from the industrial sector would be below the 1990 level in 2000, although the lack of commitment by the freight transport

Seven concrete measures are defined:

- Preparing 'industry-wise [*sic*] voluntary action plans'* incorporating definite goals and steps towards enhancement of energy efficiency, and periodically reviewing the progress of such actions.
- Recovery and use of heat produced as waste from cities and industries, reduction of natural energy costs, improvement of utilisation efficiency of fossil fuels from cities and industries, reduction of natural energy and the wise, effective utilisation of atomic energy.
- Improved energy efficiency through inter-industry collaboration based on the lifecycle assessment concept.
- Improved transport efficiency.
- Cooperation in coping with global warming in the residential and commercial sectors through the development of energy-saving products.
- Positive participation in 'activities implemented jointly' to transfer technology to developing countries in close cooperation with the government.
- Promotion of forest protection and reforestation projects in developing countries through business corporations themselves and the Keidanren Nature Conservation Fund.

*These 'voluntary action plans' were renamed the Keidanren Voluntary Action Plan on the Environment.

Source: Keidanren (1996).

and cement industries made the achievement of this goal slightly uncertain (Keidanren, 1996; Suwa, 1997: 41). The Action Plan was subject to annual review, and the results are publicly reported. However, this largely consists of self-review without guidelines, which was to invite criticism from environmentalists.

According to Hajime Ohta, a former councillor of Keidanren, the Keidanren Voluntary Action Plan was a strategy to pre-empt government action in strengthening energy efficiency policy. Anticipating a wave of enthusiasm for environmental concerns in the run-up to COP3, which Japan hosted, Japanese industry tried to avoid the costs of carbon dioxide reduction. It also wanted to keep government interference to a minimum (Ohta, 2000).

The New Earth 21: technological breakthrough

In June 1990 a 100-year earth regeneration programme, 'New Earth 21', was agreed upon by the Ministerial Council on Global Environmental Protection alongside the Action Programme. New Earth 21 enshrined MITI's belief in technological breakthroughs as a fundamental solution to the problems of the environment and the economy. New Earth 21 presented a long-term comprehensive plan for international cooperation on technology transfer and development in the energy/environment sphere. The aim was to reduce GHG emissions by 60 per cent in 2100, as called for by the IPCC.

After having formally proposed New Earth 21 to the world at the G7 summit in Houston in July 1990, MITI, in collaboration with industry, set up the Research Institute of Innovative Technology for the Earth as a primary organisation to implement R&D on innovative environmental technologies. Around 100 researchers were brought together from industry, together with (former) officials and academics, and MITI won some ¥6 billion for its first annual budget (Nakamura and Toyonaga, 1991). Industry acted in concert with MITI's initiative: Keidanren announced its intention to promote technological R&D, especially in the area of carbon dioxide disposal and fixation (Keidanren, 1990); the electricity industry began its own R&D work on carbon dioxide recovery, disposal and fixation technologies; the steel industry launched a programme to transfer energy efficiency technologies (*Yomiuri Shimbun*, 11 January 1990); and Mitsubishi Motors collaborated with Mercedes-Benz on energy/environment technologies for vehicles (*Yomiuri Shimbun*, 16 May 1990).

In order to advance specific strategies for New Earth 21, MITI further proposed the Technology Renaissance for Environment and Energy at the Tokyo G7 summit meeting in 1993. This led to the establishment of the Climate Technology Initiative, at COP1, as a formal body of the International Energy Agency (IEA), to accelerate the development, application and diffusion of climate technologies.

The social approach: lifestyle issues

In order to 'complement' the technological breakthroughs, the 'social breakthrough' was proposed by the EA's study group (Nishioka and Morita, 1992). This policy called for changes in social infrastructure, institutions and lifestyles. The problem of energy-intensive lifestyles was increasingly recognised, and the EA, MITI and the MOT strengthened their campaigns to raise awareness of the carbon dioxide problem and energy use. According to an assessment by the Central Environment Council, however, they were insufficient both in quality and in quantity (EA, 1997b: 254).

One important move on this front was MITI's attempt to introduce a summer daylight saving system or 'summer time' in Japan. MITI's argument was that the efficient use of daylight would save energy and that the introduction of summer time would also stimulate a nationwide review of lifestyles. With the FCCC coming into effect, MITI began to form an advocacy coalition with legislators to pass the Summer Daylight Saving System Bill. The first attempt was thwarted by the political turbulence caused by the rise and wane of the non-LDP coalition government between 1993 and 1994. In 1995, at MITI's initiative, a cross-party group on summer time, established by legislators in the House of Councillors, formulated a draft bill with the intention to submit it to the Diet in May 1996.

Public opinion over summer time was divided, however. The idea had been unpopular after its first four-year trial in the Occupation era, after the war, not least because it resulted in longer working hours. With MITI's efforts to build a broad coalition, public opinion was changing favourably and some industries began to see potential benefits from the summer time system. However, most supporters were passive and there remained strong opposition, especially from the construction and haulage trade unions. Outside MITI, there was little support for the bill among government ministries. In the end, the legislators could not win enough support and had to shelve the bill.

Environment taxes

The debate over an environment or carbon tax in Japan was stimulated by the growing interest in the concept at the international level in the early 1990s. In 1990, the EA and the MOF started to examine the idea of environmental taxation, focusing on a carbon tax, with the aim of introducing it in 1992 (*Nihon Keizai Shimbun*, 9 September 1990; *Mainichi Shimbun*, 2 June 1991). During 1991, with Takeshita's influence, some notable Liberal Democrats, including the then Finance Minister, Hashimoto, were also showing an interest in the idea (Schreurs, 2002: 170–3). However, the nascent idea of an environment tax became embroiled in the controversy over the extension of the temporary increases (to last one year) in taxes on oil and gas and in corporation tax, which were introduced in 1991 with the aim of financing the UN's involvement in the Gulf War. Some Liberal Democrats, MITI and the MOF separately considered the extension of these taxes, with the aim of using part of the accrued revenue for carbon dioxide reduction measures, other energy/environment measures or international environmental contributions. However, given opposition from industry (especially the oil industry) and the lack of consensus within the LDP, government failed to introduce any of them.

Nevertheless, as the Earth Summit in 1992 approached, Takeshita resumed the debate on an environment tax. He proposed introducing one as a voluntary contribution scheme, the implication being that this would be a new financial contribution from industry. The EA, the MOF, MITI and the MOFA, in response to this initiative, set up a liaison committee to coordinate the different opinions over an environment tax, in particular, a carbon tax (*Yomiuri Shimbun*, 15 April 1992). It should be noted, however, that Takeshita was in favour of an environment tax to finance international environmental contributions, and did not necessarily support a carbon tax.

The growing interest of politicians and ministries in the idea provoked strong opposition from industry. The Japan Iron and Steel Federation and the Japanese Federation of Iron and Steel Workers' Unions were the first formally to express their opposition to an environment tax, specifically a carbon tax (*Mainichi Shimbun*, 14 June 1991). The oil industry also strongly opposed it. In April 1992 the four peak associations pressured top Liberal Democrats not to introduce a carbon tax. With an election to the House of Councillors approaching in July, the LDP soon publicly announced that it had no intention of introducing any environment tax in the near future (*Asahi Shimbun*, 17 April 1992), and instead the peak associations were to contribute to environmental funding (*Asahi Shimbun Evening*, 8 May 1992).

The controversy over the environment tax continued as deliberations on a new Basic Environment Law started in July 1992 (this new Basic Law was a Japanese response to the Earth Summit in

June 1992 on sustainable development). In September, the EA's study group on an economic system to arrest global warming reported that some form of economic instrument would be very effective in reducing carbon dioxide emissions and financing necessary measures (Study Group on the Economic System to Arrest Global Warming, 1992). MITI, on the other hand, drew the conclusion from its two councils that environment taxation could not be made to work. In contrast to the EA's report, which argued that it was possible to design a cost-effective economic instrument, MITI's councils emphasised the likely negative effects on the economy and pointed to the lack of public understanding of the need for an environment/carbon tax (Industrial Structure Council, Advisory Committee for Energy, and Industrial Technology Council, 1992). The clause concerning economic instruments in the EA's preliminary draft of the Basic Environment Bill fuelled opposition from industry and MITI. During the final negotiations between the EA and MITI the clause was watered down and the use of economic instruments was made conditional on several requirements. First, the burden of any economic instruments should be 'equitable' (meaning they should not generate particular losers). Second, if taxes are necessary, the government must acquire positive consent from the public. Third, when such taxes are for global environmental protection, international collaboration should be considered. Last, the environmental effectiveness of a tax must outweigh its cost to the economy (Basic Environment Law, article 22, para. 2). For MITI, these conditions virtually excluded the possibility of a carbon tax being introduced.

This outcome may not be surprising, given the objections to a new economic burden on the public from virtually all political parties, consumer groups and even some environmental groups, let alone industry (Japan Environment Council, 1994).[5] Nevertheless, the controversy raised awareness of the issue, and the government's Tax Commission for the first time referred to environment taxation in its report on tax reform for the fiscal year 1993, arguing for further research on it.

Thereafter, the EA continued to study environmental taxation and various opinion surveys indicated increasing (if reluctant) acceptance of a carbon tax from *individual* major companies (EA, 1996; Sawa, 1997: 143).[6] Nevertheless, strong opposition from *industrial associations*, in particular Keidanren, persisted.

Keidanren also made a joint declaration with the Federation of German Industry maintaining that an environment tax would damage the economy but produce little benefit for the environment, while emphasising the importance of voluntary action (Henkel and Toyoda, 1996). MITI's opposition also continued. In July 1996, the EA's study group concluded that a low level of carbon tax with hypothecated provision for energy efficiency would reduce carbon dioxide emissions slightly by 2000 and by 3 per cent by 2010, with negligible effects on the economy. Against this, MITI issued a report in March 1997 which formally rejected the EA's proposal, by raising doubts about its environmental effectiveness and pointing to the existing taxes on oil and gas, part of the revenues from which were used for energy efficiency and 'new energy' measures (Subcommittee on Global Environment, Industrial Structure Council, 1997).

It should be noted that despite consistent opposition to an environment tax, MITI considered increasing the taxes on oil and gas and reforming existing energy taxation to respond the carbon dioxide reduction imperative. The idea was, however, abandoned because of the prolonged economic recession Japan was then experiencing (*Nikkan Kogyo Shimbun*, 28 November 1992; *Nihon Keizai Shimbun*, 10 March 1993). When MITI opposed the EA's environment tax, it was concerned not only about its effect on the economy and industry, but also about green interference in its jurisdiction.

Conclusions

Some corporatist institutional characteristics – the close government–industry relations and consensus-based policy-making – can be detected in the Japanese policy response to global warming. The government used carrots such as various financial incentives and soft, flexible instruments in the expectation of self-regulation and compliance rather than heavy-handed government regulation. Policy concertation was seen in the New Earth 21 initiative, the establishment of the Research Institute of Innovative Technology for the Earth and various private investments in technological R&D and voluntary action by industry. Industrial associations played a key role here. MITI relied on these associations in mobilising cooperation from individual companies for improved energy efficiency. Most importantly, Keidanren, whatever the motives, orchestrated industrial action to cooperate with the government's effort to reduce carbon dioxide emissions. The corporatist, close government–industry relations, the relatively well integrated business sector and the practice of policy concertation enabled the government to shape its policy strategy. This strategy was characterised by a collaborative, technological approach.

The apparently pluralistic process of target-setting also exhibited characteristics associated with consensus corporatism. The EA and MITI had different 'philosophies' and represented conflicting economic and environmental interests. Still, the norm of consensus policy-making prevailed. Without strong political leadership the decision-making was inevitably slow. The ambiguous outcomes also indicate the non-'winner takes all' mentality in Japanese policy-making. Power was shared by the EA and MITI, which represented conflicting interests. However, in the end MITI overcame the objections of the EA in both target decisions, for 2000 and 2010.

Economic interests were, indeed, an important factor shaping Japanese policy. Industry, while collaborating with government, also defended through various channels its interests against some of the government's policy initiatives. A case in point is a carbon tax. Corporatism's collaborative bargaining may give insight here; however, what also appears to be important in the politics of a carbon tax was the self-interest of MITI and the LDP. For MITI, the EA's carbon tax represented 'green interference' with its energy jurisdiction. Also, in the run-up to an election, opposition to a carbon tax was rational behaviour for vote-maximising politicians. Here, the issue-based approach, which casts light on the importance of self-interests, may be useful in explaining policy developments. The issue-based approach further suggests a serious problem over collective action that is inherent in a carbon tax. With little support for the tax and no clear winners from it, the most affected groups, namely energy-intensive industries, were effective in defending their interests. A not dissimilar case was the failure to introduce a summer daylight saving system. Small groups that stood to lose from the introduction of summer time thwarted MITI's attempt to introduce it.

The institution-based account of target-setting should be balanced by an account of the effects of an individual policy catalyst and international pressures. At the first policy change on target-setting, Prime Minister Takeshita mobilised support from a number of influential Liberal Democrats for global environmental issues, thus bringing about a political climate of unprecedented favourability to the EA. There were also international pressures for policy change emanating from the epistemic community, which had reached a (virtual) consensus on the likelihood of global warming, and from the decision to start international negotiations on the FCCC, as well as from the fact that Japan was due to host COP3. These international pressures combined with the domestic political climate to accelerate policy change. In the run-up to COP3, Prime Minister Hashimoto directly accelerated the decision-making process and gave support to the EA on targets. However, Hashimoto kept the decision-making process closed to any politicisation. Moreover, although environmentally sympathetic politicians and political parties acted on the issue, their move to 'get on board' was too late. Despite Hashimoto's implicit support for the EA, therefore, the political climate was not as favourable when the EA was negotiating targets for 2000. Consequently, the imbalance of political power between the EA and MITI left a clearer imprint on the second target.

To summarise the main analytical conclusion: the institutional approach gives a good insight into policy contents. The politics of target-setting and the mix of policy concertation and the importance of economic interests in shaping policy also indicate that the speed of policy change and policy stringency can be at least partly explained by the institutional approach. However, the institutional analysis is certainly not sufficient. In the case of target-setting, it has to be balanced by taking into account the role of individual policy catalysts and international influences. The issue-based approach also gives some insight into political outcomes. The role of institutions and interests in explaining policy will be explored when Japanese is compared with British policy. Meanwhile, the next chapter will give a further view on the relationships between the two approaches, which looks at policy integration in Japan.

Notes

1 Before Takeshita took up global environmental issues, Japan had proved rather indifferent to the growing international criticism

levelled against the Japanese government and industry about their increasingly environmentally threatening economic activities.

2   Article 9 of the Constitution of Japan requires the Japanese people to 'renounce war as a sovereign right of the nation and the threat or use of force as means of settling international disputes'.

3   In the end, because of objections mainly from Australia, oil-producing countries and Russia, the Geneva Declaration was not 'adopted' but 'taken note' of.

4   It was found that there was not enough solar light shed on vending machines, as they are usually located in the shade (interview with Kenji Fujita, 2000).

5   Komeito (or the Clean Government Party, an offshoot of Japan's largest lay Buddhist organisation) was a notable exception in its support for the use of environment taxes.

6   For example, in 1992, 39 per cent of major firms were in favour of environment taxes and 38 per cent were opposed, whereas in 1995 46 per cent supported the taxes and 38 per cent were opposed.

6

Co-optation and exclusion: controlled policy integration in Japan

To the extent that the problem of global warming arises from existing socio-economic activities, tackling it will entail an institutional metamorphosis towards a more sustainable form of socio-economic system. This will require a realignment of broad policy goals, which itself may require changes in policy-making institutions. Such changes have been referred to as policy integration, which is the theme of this chapter.

The integration of environmental concerns into general economic policy in Japan started around the time of the Earth Summit in 1992. The 1992 *Five-Year Plan for Livelihood Great Power – Sharing a Better Quality of Life Around the Globe* (Economic Planning Agency, 1992) argued for the need to establish harmony between a demand-led economy and the environment, and the need to pursue a high quality of life rather than a high quantity of production. The Economic and Social Plan for Structural Reform agreed on by the Cabinet in December 1995 incorporated a good number of environmental ideas, including the creation of a sustainable and zero-waste economic society, and the need to rethink prevailing lifestyles, based as they were on mass production, mass consumption and mass disposal. In 1994, the Basic Environmental Plan under the Basic Environment Law set forth a new framework for energy and transport policies, although it did not set numerical targets (EA, 1994a: 73–6).

This chapter looks at policy integration specifically in the areas of global warming and energy and transportation. There are inevitably overlaps with the previous chapter, but the primary concern of this chapter is whether and how global warming policy *caused* policy and institutional developments, as opposed to the previous chapter's focus on policy as an *effect*.

The energy policy background

The most important determinant of Japanese energy policy has been the frail supply structure. While currently Japan is the world's fourth largest consumer of primary energy, more than 80 per cent of this energy is imported. Before the first oil crisis hit the world, Japan depended for 72 per cent of its primary energy supply on oil. The two oil crises drove home to Japan its need to secure a stable supply of energy. ANRE, which was established within MITI just before the first oil crisis, placed strong emphasis on enhancing energy security.

There were two further characteristics of traditional Japanese energy policy, linked to this prime objective, that should be noted before policy integration is analysed. First, the Japanese energy industry was firmly protected from excessive competition through various regulations, while being obliged to ensure a stable supply of their products. The typical example was the electricity industry. Until the mid-1990s ten private, vertically integrated electricity supply companies were granted local monopolies. Price regulation was designed to cover necessary costs and a fair rate of return on investment, and thus MITI and the LDP had the final say over the rates. A series of protective measures inevitably made the industry inefficient, and Japanese electricity prices were some of the highest in the world. The second characteristic was the existence of systematic energy conservation measures. After the oil crises, MITI posited energy conservation as a key to enhancing energy security. However, it did not necessarily mean that Japanese policy-makers pursued demand-side measures. On the contrary, they essentially aimed to meet the predicted maximum demand, which was, in turn, expected to rise as the economy expanded.

It is often said that this supply-side approach was an inevitable outcome of the existing pattern of energy policy-making. The most important arena for overall energy policy-making is the ACE, which was, in effect, dominated by representatives of the energy producers, major energy consumers and former MITI bureaucrats or representatives of MITI's affiliated organisations. Until the Cabinet decision of September 1995 to disclose, in principle, all the minutes and materials of advisory councils and committees, it had been extremely difficult to conduct an in-depth public review of the ACE's policy for those who did not have access to the ACE, as was the case for environmental, citizens' and consumer groups (People's Research Institute on Energy and Environment, 1994: 10–11; Japan Federation of Bar Associations, 1999: 33–40, 56–8). When policy was made behind closed doors by economic bureaucrats, drawing on information from major energy-consuming industries and the electricity industry, which had an obligation to ensure a stable supply of energy, the resultant policy inevitably had a supply-side orientation (Japan Federation of Bar Associations, 1999: 37, 47).

When energy policy-makers faced the imperative of policy integration in the 1990s, there were at least two issues in conjunction with which the integration process had to be examined. One was the growing fear of electricity crisis. Accelerating peak demand for electricity, especially during the summer, and the difficulties in finding sites for new power plants and the long lead time before they came on stream, had pressed the government to take action to reduce or at least shift the peak demand.[1] The other issue was deregulation of the energy industry, especially electricity markets. Reforms to electricity price regulation, aimed at reducing electricity prices, in particular, had important implications for policy integration.

Energy efficiency policy: in search of a zero-sum solution

Energy policy-makers responded fairly quickly to the looming political concern about global warming in the late 1980s. In October 1988 a private advisory body was established for the Director-General of ANRE to discuss a long-term energy policy in response to the rise of global environmental issues. The report of that advisory body led to a comprehensive review of existing energy policy, including the Energy Outlook of May 1990 (see Chapter 5). At the end of 1989, MITI convened the ACE's Coordination Subcommittee for the first time in fourteen years and, jointly with the Energy Supply and Demand Subcommittee, started to deliberate on a mid- to long-term energy policy and to review the Energy Outlook. The ACE's Subcommittees on Energy Conservation, Alternative Energy, Nuclear Power, and Coal were also convened to review and discuss detailed policy in their fields. One of the key figures in the ACE, Yoichi Kaya, was deeply involved in the international politics of global warming and was also a member of the IPCC's WG-III, which examined possible responses. Kaya communicated extensively with the MITI officials responsible for drawing up the Energy Outlook. According to Kaya, within the context in which the global warming problem loomed large in both international and domestic politics, there was a broad consensus among the members of the ACE that limiting emissions of carbon dioxide was a priority in drawing up the next Energy Outlook (interview with Yoichi Kaya, 2000).

The outcome of the ACE's deliberations was a milestone. For the first time, environmental objectives were posited as a guiding principle of the Energy Outlook, alongside the traditional emphasis on stability of energy supply (interview with Yoichi Kaya, 2000). In contrast to the past Energy Outlooks, which had always overestimated energy demand, the new Outlook looked at the need to constrain demand. Compared with a 4.8 per cent annual increase in energy demand in 1987 and 5.7 per cent in 1988, the ACE required Japan to contain the growth in energy demand to within 1.6 per cent annually between 1988 and 2000, leading to a 6 per cent reduction in energy demand in 2000 from the projected business-as-usual level (ANRE, 1990).

While energy policy-makers presented the new energy efficiency targets as 'very ambitious', not all were satisfied with them, notably the EA, which engaged in heated negotiations with MITI over the stabilisation target. Also, Greenpeace Japan and the People's Research Institute on Energy and Environment presented alternatives to the ACE's Energy Outlook, and demonstrated that it would be possible even to reduce energy demand by 0.4 to 2.4 per cent annually between 1990 and 2010 (People's Research Institute on Energy and Environment, 1994).

The new Energy Outlook soon came to be seen as 'unrealistic', especially given MITI's approach to energy efficiency, which largely relied on voluntary action. Reviewing the 1990 Energy Outlook, Toyoaki Ikuta, a key figure in energy policy-making and the then chairman of the Energy Supply and Demand Subcommittee, warned that without a robust energy policy it would be difficult to achieve the energy targets. In September 1991, MITI convened the ACE's Energy Conservation Subcommittee to look at how to strengthen existing energy efficiency measures, in particular the Energy Conservation Law. However, amid the bubble economy, industrial members of the ACE and also the industrial sectoral bureaus of MITI were resistant to the idea of strengthening the

Law. Having failed to obtain their consent, the energy policy-makers' hopes of amending the Energy Conservation Law in the first half of 1992 were shattered. The Energy Conservation Subcommittee suspended its deliberations in April 1992, and decided to wait to see the outcome of the Earth Summit in June 1992 (*Enerugî to Kankyô*, 12 March 1992; Fukushima, 1995).

Soon after the suspension of these deliberations on energy efficiency, MITI produced a new framework policy on energy and the environment. It presented as a new principal objective of energy policy the simultaneous attainment of the three Es – economic growth, energy security and environmental protection. A similar idea had already been advanced by Keidanren (1989) in an opinion paper on global warming. In this sense industry's view formed a central element of MITI's new policy framework. Government and industry were ready for concerted action towards a non-zero-sum energy/environment policy. Just after the Earth Summit, MITI resumed its deliberations by establishing a joint council composed of the main members of the ACE and MITI's Industrial Structure Council and Industrial Technology Council. During the deliberations, MITI, which had been ambiguous regarding its view on whether the carbon dioxide stabilisation target was attainable, for the first time indicated that it was, and went so far as to suggest that the carbon dioxide problem could be tackled in a positive-sum manner, and that the three Es could be attained with realistic policy measures such as revisions to the Energy Conservation Law and the further use of incentives (*Enerugî to Kankyô*, 17 September 1992).

Although resistance from industry to a stringent energy efficiency regulation remained, in October 1992 MITI produced its New Strategy for Energy Conservation and Environmental Protection, the main plank of which was strengthening of the Energy Conservation Law. The aim was to reduce energy demand in all sectors by 7.5 per cent on average in 2000 from the business-as-usual level (*Nihon Keizai Shimbun*, 28 October 1992; OECD, 1994: 147). Building on deliberations at the joint council, which emphasised the importance of using 'carrots' rather than 'sticks', MITI prepared new incentives. These policies and measures were, in a sense, a pre-emption of the EA's move at the time. The EA intended to develop and implement global warming policy through legislation, namely the Basic Environment Law which it was

hammering out. With its New Strategy, MITI forestalled the introduction of measures to reduce carbon dioxide emissions before the
EA could strengthen its general competence over global warming
policy. Allegedly MITI's original plan to include an environmental
element in its proposed revisions of the Energy Conservation Law
was opposed by the EA, which was concerned about who had
jurisdiction over global warming policy.

As explained in Chapter 5, in 1993 a package of energy efficiency policies, centred on the amended Energy Conservation Law,
was introduced. However, it soon proved insufficient to achieve
the carbon dioxide stabilisation target, and MITI started to engage
in 'number juggling' to make its energy forecast and the stabilisation target consistent with each other. In 1994 the ACE required
Japan to contain the growth in energy demand to within 1 per
cent annually until 2000. In fact, energy consumption reached
this forecast level for 2000 as early as 1995, which left the ACE
requiring no further increase in energy demand after 1995. Many
commentators, including major members of the ACE, criticised
this target as unrealistic (interview with Yoichi Kaya, 2000). In
March 1997, MITI outlined new and strengthened measures for
energy efficiency (Council of Cabinet Ministers for the Promotion
of Comprehensive Energy Policy, 1997). Apart from the introduction of fuel efficiency standards for diesel vehicles, they were
basically a more stringent application of the Energy Conservation
Law and an expansion of financial incentives, which were hardly
commensurate with the ambitious zero-growth target.

Electricity prices and demand
According to the ACE, the main reason for the lack of progress
in energy efficiency in the 1990s was lower energy prices, which
inevitably discouraged investment in energy efficiency; moreover,
energy-intensive business and lifestyles were brought about by
an affluent economy, as symbolised by the prevalence of various
electrical appliances and cars (ANRE, 1997: 20–1). The growth in
total energy demand had been decoupled from economic growth
after the oil crisis in the 1970s (see Chapter 5); however, the
growth in electricity demand virtually matched economic growth.
After 1993, not least as a result of the political pressure to reduce
utility rates, the increase in electricity demand outpaced economic
growth (see Figure 6.1).

Figure 6.1 *Energy, electricity, electricity prices and GDP, 1970–2000 (1973 = 100).*

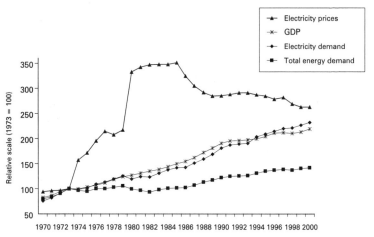

Source: Energy Conservation Centre (2003).

In 1994, in response to mounting socio-political pressure to deregulate the electricity markets and reduce electricity prices, the Electricity Utility Industry Council considered major reforms to the pricing of electricity. During these deliberations a great deal of time was spent in obtaining the consent of the electricity companies to a reduction in prices while at the same time trying to ensure the introduction of an incentive mechanism to encourage a rationalisation at least of electricity usage. In the event, the final report did not even refer to the problem of global warming (Rate System Committee, Electricity Utility Industry Council, 1995). Nevertheless, the new pricing system was presented as conducive to a reduction in carbon dioxide emissions, because it made it much easier for electricity supply companies to adjust their rates to discourage consumption at periods of high demand, thus levelling the electricity load. MITI and the power companies maintained that such load-levelling was a key not only to ensuring security of supply and lower prices, but also to reducing emissions – a seemingly win–win policy. This argument, however, obscured the crucial issue of the possible effect of lower electricity prices on overall electricity demand. The issue was kept off the agenda.

Despite the significant environmental implications of its deliberations, there was no environmental expert on the Council, let alone a representative from an environmental NGO, although it must be conceded that the latter probably lacked the capacity to take action on the issue at the time (interview with Mie Asaoka, 2000).

New energy: is policy changing?

The problem of global warming gave a new impetus to a receding 'new energy' policy in Japan. Early in 1990 the ACE's Alternative Energy Subcommittee started to review the policy. Notable was a reference to the USA's Public Utility Regulatory Policy Act (PURPA) by some members of the Subcommittee as a way of promoting 'new energy' (on which, see Chapter 5). Heavy-handed PURPA-like legislation, which would have required electricity companies to purchase electricity generated from small, non-utility power generators, was almost a taboo for power companies in Japan (*Enerugî to Kankyô*, 18 January 1990, 16 April 1990). Outside the ACE, the PURPA was also being examined by the EA as part of global warming policy, and in 1991 the biggest opposition party, the Social Democratic Party, drew up a bill that would require electricity companies to purchase surplus electricity generated from the combustion of waste.[2] Power companies were under great pressure to do something about new energy. It was in this context that they started a voluntary scheme to purchase surplus electricity generated from renewable sources. The voluntary scheme was a direct response to the interim report issued in June 1991 by the Electricity Utility Industry Council, on which power companies of course had strong representation.

With the introduction of the voluntary purchase scheme by power companies, the Social Democratic Party dropped the bill on the grounds that its objectives had largely been achieved already. However, the idea of obliging electricity companies to purchase a certain amount of new energy stayed afloat. In its 1997 white paper the EA pointed out that although the power companies' voluntary scheme and the new bidding system for the purchase of electricity by utility companies (which was introduced after the 1995 Electricity Utility Industry Law had deregulated the industry) had expanded opportunities for new energy – specifically energy from waste – they were not sufficient, with the implication that a mandatory purchase scheme was necessary (EA, 1997a: 99–103). Similarly, environmental groups increasingly sought to have a new obligation placed on the electricity companies. MITI and the power companies, preoccupied with electricity deregulation and liberalisation and with reducing electricity prices, insisted on the need for any action to be voluntary. That said, MITI was also enthusiastic about expanding the markets for new energy. The Programme for Economic Structure Reform, drawn up by MITI and decided by the Cabinet in December 1996, for example, identified new energy and energy conservation as being among fifteen new and growth fields to be promoted.

In February 1997, with the generation of new energy well below the ambitious target set in 1994, MITI introduced the New Energy Bill, to be the first law to promote new energy. The bill was basically another attempt to strengthen the existing, voluntary approach to the construction of new energy facilities, with 'carrots' of financial incentives and a 'soft stick' of administrative guidance. In the face of the power companies' opposition and MITI's reluctance to adopt a regulatory measure to oblige power companies to purchase new energy, this approach was perhaps the main policy option left to Japan.

Nevertheless, when the Diet passed the New Energy Bill, a supplementary resolution was adopted which required the government to explore the introduction of a mandatory purchase scheme. A more notable development was the very unusual degree of cooperation between industry and an environmental group. People's Forum 2001 and the Tokyo Electric Power Company launched a collaborative subsidy programme for the installation of residential photovoltaic power systems in 1997. By linking the programme with MITI's subsidy scheme, this three-year programme saw the installation of these systems in 263 households (*Gifu Shimbun*, 7 June 2001). An environmental group joined in the concerted action between government and industry to promote photovoltaic power.

New energy and nuclear energy
It is often pointed out that the Japanese government's blind faith in nuclear energy has hindered the healthy growth of new energy. How the strong commitment to nuclear energy has overshadowed support for new energy has been clearly reflected in the

budget. In 1996, ¥500 billion was allocated for nuclear development and promotion, while only ¥15 billion in total was given to
new energy (Climate Network Europe and United States Climate
Action Network, 1996).

Indeed, nuclear energy policy in Japan is both systematic and
institutionalised. The Atomic Energy Basic Law was enacted in
1955, in accordance with which the Long-Term Programme for
Development and Utilisation of Nuclear Energy is formulated
every five years by the Atomic Energy Commission. In 1956
the Science and Technology Agency was established chiefly to
administer R&D in nuclear energy.[3] There are also two public
corporations for nuclear research and operation. Finance for
nuclear promotion is secured through the Special Account for the
Promotion of Development of Electric Power Sources and raised
via a tax on electricity use. More than half the revenue raised
from this tax was used for nuclear R&D and for finding sites for
nuclear facilities. Throughout the postwar era, thus, the government has built centralised institutions for nuclear energy planning,
finance, R&D, siting and operation, without establishing effective
institutional mechanisms for public scrutiny (Takei, 1995).

There are interlocking interests among the industries involved
in the construction and operation of nuclear power stations,
and these extend to politicians and bureaucrats. Construction
of nuclear power stations brings significant economic profits to
various industries, and notably to companies involved in the
production and supply of heavy electrical equipment and major
construction companies. The existing electricity pricing system
allows electricity companies to increase profits in proportion to
their investment in expensive nuclear power facilities. Some local
authorities are heavily dependent on the subsidies from the central
government which a nuclear power plant brings. Thus, for politicians, nuclear power stations represent an excellent 'pork barrel'.
For MITI officials, nuclear energy brings discretionary power
emanating from regulations and money from the tax on electricity
consumption. The nuclear power industry, in which a number of
quasi-governmental and non-governmental bodies operate, also
provides MITI officials with lucrative placements after retirement.

Compare the nuclear power policy – institutionally and politically firmly in place – with that on new energy. MITI and the
electricity companies openly distrust the reliability and viability of

new energy as a substantial source of energy. For them, new energy is subject to weather, and hence unstable, and this instability would place a large burden on electricity companies' power systems if new energy were introduced on a large scale. They also consider new energy expensive and inefficient in terms of space compared with large-scale, centralised power plants (interview with Masanobu Hasegawa, 2000). While recognising that there is a good market opportunity for new energy technologies and that there is a need to increase the supplies of new energy (interview with Yoichi Kaya, 2000), neither MITI nor the electricity companies are prepared to give it more than a marginal role in electricity generation. Indeed, in its 1993 white paper on energy, MITI warned that the growth in dispersed energy sources should not hinder the development of nuclear energy (ANRE, 1993).

The EA's stance on nuclear energy differed from that of MITI and was more closely aligned with that of environmental groups (see Table 6.1). Although the EA officially advocated the national policy of promoting nuclear energy, it apparently did not intend to rely on it as a key to reducing emissions of carbon dioxide. In two early studies, concluded in 1990 and 1992, the EA demonstrated that stabilisation of carbon dioxide emissions in 2000 would be attainable without additional nuclear plants to those under construction. Given the increasing difficulties in finding sites for them, as a result of growing anti-nuclear sentiments, the EA's policy of having no further nuclear power plants was perhaps more realistic than MITI's plan.

MITI was firm in its stance on nuclear energy, even after a near crisis in policy was triggered by a series of serious accidents and cover-ups at nuclear facilities from 1995. At the deliberations of the ACE in 1996, it presented a study showing that without a further fifty new nuclear power plants, even the most stringent energy

Table 6.1 *Differences in 1990 between the EA and MITI in the projected contribution (in millions of kW) of nuclear energy*

	1988 figure	2000 projection	2010 projection
MITI	28.87 (with 12.85	50.5	72.5
EA	under construction)	41.55	41.55

Source: ANRE (1990); EA (1990c).

efficiency and new energy policies would not be able to stabilise emissions of carbon dioxide by 2030 (ANRE, 1997: 147–69). Although MITI had long implied that nuclear energy was key to solving the global warming problem, this was the first time that it made the argument so explicitly.

Energy taxation and finance, coal and global warming: lack of environmental debate

The analysis of Japanese policy integration cannot be completed without examining energy taxation. As shown in Table 6.2, all energy taxes were earmarked for specific purposes. As far as energy policy was concerned, the structure of energy taxation and expenditure generally reflected the priority placed on reducing the nation's dependency on oil and achieving security of energy supply. After the appearance of the problem of global warming on the national agenda, major expenditures for energy efficiency and new energy from the Energy Special Accounts grew (see Table 6.3). In 1993 the item 'energy efficiency and new energy' was created in the Coal and Petrol Special Account. On the other hand, coal, which is responsible for the most emissions of carbon dioxide, was most favoured among fossil fuels. Coal was, until 2003, the only fossil fuel that was not subject to an energy tax.[4]

The exemption of coal from energy taxes was originally to protect the domestic coal industry. By 1990 domestic coal accounted for only about 10 per cent of the total coal used in Japan; however, this coal protection measure appears to have given a twist to policy integration. One plank of the coal protection policy was a deal under which major coal users, such as the electricity, steel and cement industries, were required to buy assigned amounts of domestic coal, which was two to four times more expensive than imported coal. By 1990, although most coal consumers were freed from this obligation, largely for economic reasons, the electricity companies had to continue to bear the cost, under strong pressure from MITI and the LDP (Lesbirel, 1991). When the government was forcing coal consumers to shoulder the cost of its coal protection policy, it would have been politically difficult then to introduce a tax on coal.

This tax structure encouraging the use of coal was also closely tied with the objective of enhancing energy security. Being widely

Table 6.2 *Energy taxation and expenditure, fiscal year 2004*

Budget used	Source of taxation	Amount raised (yen)	Expenditure	Total amount allocated
Special Accounts for Coal, Petroleum and the More Sophisticated Structure of Demand and Supply of Energy Policies	Customs duties on crude oil products[a]	38 billion	Coal policy	54.1 billion[a]
	Petroleum tax (currently petroleum and coal tax)	477 billion	Oil policy Alternative energy, energy efficiency and environmental policy	367.8 billion 256.3 billion
Road Improvement Special Accounts	Petroleum gas tax Gasoline taxes Diesel oil delivery tax	28 billion 3,139.7 billion 1,075 billion	Road building and improvement	4,177 billion
Airport Construction and Improvement Special Account	Aircraft fuel tax	106.4 billion	Airport building and maintenance	472 billion
Special Accounts for Electric Power Development	Power sources development tax	359.3 billion	Promotion of power station siting Diversity of electricity supply	257.7 billion 245.6 billion

[a] The current account for coal is a temporary measure (to last up to financial year 2006), reflecting the official termination of the coal rationalisation programme in 2001. Small-scale protection still lingers on, however.
Source: Ministry of Environment (2004b); Petroleum Association of Japan, *Annual Review 2003* (2004); Ministry of Economy, Trade and Industry, www.meti.go.jp.

Table 6.3 *Growth in major budgets for energy efficiency and new energy in the Energy Special Accounts (billion yen)*

	1992→93	1996→97	2002→3	Increase as ratio of 2003 value/ 1992 value
Energy efficiency measures	27.3→49.0 (+79%)	61.0→77.3 (+26.7%)	121.9→131.8 (+8%)	4.8
Promotion and dissemination of alternative energy sources to oil	5.4→6.8 (+24.5%)	10.3→13.9 (+35.4%)	65.3→76.2 (+16.7%)	14.1
Development and promotion of clean energy	38.4→38.7 (+0.8%)	48.1→49.1 (+1.9%)	82.7→87.8 (+6%)	2.3
Total growth in the Energy Special Accounts	1,052.1→1,091.7 (+3.8%)	1,191.7→1,242.0 (+4.2%)	1,112.2→1,118.1 (+0.53%)	1.1

These expenditures do not include those for the NEDO.
Source: Ministry of Economy, Trade and Industry, www.meti.go.jp; Petroleum Association of Japan, various years.

Figure 6.2 *Primary energy sources for electricity generation, 1970–2002.*

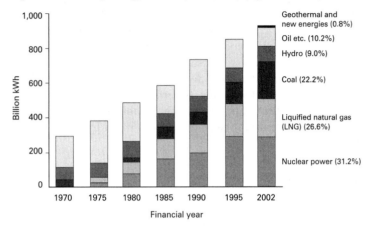

Source: ANRE (2003).

located in politically stable countries, and cheap, coal was considered a good alternative to oil (ANRE, 1994a; EA, 1997a: 71). Although energy policy-makers recognised the environmental constraints on the use of coal, they did not explicitly encourage a shift away from it. The extent to which the use of coal was favoured for electricity generation can be seen in Figure 6.2. Facing the combined problem of finding sites for nuclear power stations and rapid growth in electricity demand since the end of 1980s, the government let the electricity companies opt for coal in order to meet demand. With energy policy-makers reluctant to let the contribution of LNG to overall energy output grow further than the existing level of about 20 per cent, coal was used as the main substitute for nuclear energy in the light of energy security. Partial liberalisation of the electricity markets also encouraged newcomers to go for cheap coal power generation. After 1990, coal became the fastest-growing energy source for electricity generation.

This policy was not, of course, without its objectors. In 1992 a coal-fired power plant was frozen for the first time, for reasons involving concerns about global warming. Local protesters, anxious about the effects of a new power plant on the landscape, worked with global environmental NGOs. In the same year, against a ten-year power development plan which forecast the fastest growth in the use of coal, the EA claimed there was a need to strengthen energy efficiency policy and to increase the use of dispersed energy sources, in the light of the imperative to reduce emissions of carbon dioxide. However, these objections did not affect the general trend favouring coal. Although the power development plan laid out measures for the efficient use of coal, they were basically formulated without reference to the stabilisation targets for carbon dioxide (*Enerugî to Kankyô*, 26 March 1992, 23 July 1992). The traditional 'stable supply' approach marginalised an important issue that has to be addressed to tackle global warming.

Transport and global warming

The transport sector contributes about 20 per cent of total emissions of carbon dioxide and this is expected to increase. Nearly 90 per cent of emissions in the transport sector are from road transport, and emissions from passenger cars account for about half the total emissions in the transport sector (Figure 6.3); this figure has increased rapidly in line with longer in distances travelled and people's increasing general preference for larger cars.

Japanese transport policy-making is decentralised. Although the MOT has prime responsibility for transportation administration, including infrastructure improvements and the rail, marine, air and road transport industries, the MOC's Road Bureau is in charge of road-building. The car manufacturing industry is under MITI's jurisdiction. Traffic regulations are the National Police Agency's responsibility. The EA has a role in environment-related transportation issues.

When the global warming problem appeared on the political agenda, heavy road traffic was already a serious social, economic and environmental problem. The growing number of road accidents was even likened to a 'traffic war'. The annual cost of congestion was said to be ¥12 trillion or about 2–3 per cent of GDP (Mizutani, 1996: 161). The government repeatedly failed to set environmental standards for the levels of nitrogen oxides and small particulate materials, or for noise and vibration in urban areas.

The MOT quickly responded to the global warming issue by launching scientific research on climate change through its Meteorological Agency and by creating a new section for dealing

Figure 6.3 *Breakdown of carbon dioxide emissions in the transport sector, 2001.*

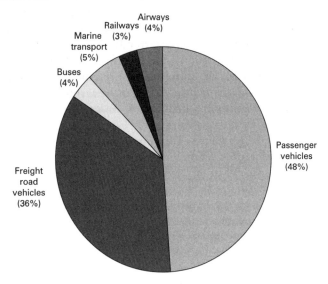

Source: Road Bureau, Ministry of Construction (2003).

with global environmental issues. Soon after the Basic Environment Law was enacted in December 1993, the MOT and MOC separately decided an intra-ministerial guideline for the integration of environmental considerations into their own administration, even before the EA formulated the Basic Environment Plan. Nevertheless, compared with MITI, the MOT and MOC were slow to act to reduce emissions of carbon dioxide.

Improving fuel efficiency
The average fuel efficiency of passenger cars peaked in 1982 and then fell. After the abolition of excise tax and revisions to the tax on vehicle possession in 1989, the gap in tax burden between large and small cars was significantly narrowed, and this stimulated demand for larger and therefore more fuel inefficient cars. Between 1988 and 2000 the proportion of cars with an engine size of over 2000cc increased from 4 per cent to 35 per cent (EA, 1994b: 65; Kashima, 2003: 11).

Facing deteriorating fuel efficiency and the problem of global warming, MITI and the MOT reviewed and strengthened the fuel efficiency standards under the Energy Conservation Law. Improvements in the fuel efficiency of individual vehicles were, however, being offset by the general trend towards larger cars. The government was slow to address the problem. In 1992 a research group composed of MOT officials considered the use of economic instruments as a way to reduce emissions of carbon dioxide from road transport (Transport Economics Research Centre, 1992: 38–57). It was, however, only in 1997, in the run-up to COP3, that the MOT, through its Transport Policy Council, formally advanced the idea of greening automobile taxation as a short- to medium-term measure to influence people's preferences in relation to cars. Acknowledging the negative effects of the tax reform in 1989 on people's choice of cars, the Council proposed a restructuring of vehicle-related taxes. Notably its report also discussed the introduction of a carbon tax in the transport sector (Transport Policy Council, 1997). Detailed discussion was, however, left until after COP3. The Epilogue to the present work looks at its development, which was characterised by constraints arising from economic interests and jurisdictional competition.

Concerted action for the promotion of low-emission vehicles
One feature of the Japanese approach to fuel efficiency is the enthusiasm for the development of low-emission vehicles (LEVs). LEVs had been promoted since the 1970s in response to the oil crises and to tackle air pollution problems; however, past policies to promote LEVs had not borne fruit. From the end of the 1980s, two policy and market developments stimulated government's hitherto rather half-hearted attitude towards LEVs.

First, a detailed medium- to long-term strategic plan to disseminate LEVs was produced by the EA as a means of reducing emissions of nitrogen oxides. The EA also envisaged introducing an obligation on business to have a certain proportion of fleets comprise LEVs (*Enerugî to Kankyô*, 12 May 1988). The LDP's policy committee on the environment also supported strengthening LEV policy (*Enerugî to Kankyô*, 6 July 1989). Second, and more importantly for MITI and industry, in 1990 the US government passed the Clean Air Act Amendments, which incorporated a requirement on car manufacturers to build LEVs, and the state of California required car manufacturers, including those of Japan, to make the sale of electric vehicles account for at least 2 per cent of their total sales from 1998, rising to 10 per cent by 2003. Moreover, General Motors had already introduced an electric car to be developed for commercial production (*Japan Eco Times*, 1995).

The EA, MITI and the MOT started to reconstruct and strengthen their policy on large-scale dissemination of LEVs. MITI, for example, increased the LEV spending in its Energy Special Account by a factor of eight. The government's interest in LEVs stimulated car manufacturers, electricity companies, and the gas and steel industries to invest in LEV development (*Nihon Keizai Shimbun*, 31 March 1990). Also, with the initiative of MITI and the MOT, sixteen industrial associations representing LEV producers, users and fuel suppliers were brought together to establish the LEV Forum, in 1992, as a locus for information exchange and cooperation. In 1995 an environmentally sympathetic Liberal Democrat, Takashi Kosugi, established a subcommittee on the promotion of LEVs under the LDP's environment committee, which produced a report that urged government to strengthen measures (Subcommittee on Promoting the Spread of Low-Emission Vehicles, 1996). With this, the EA, MITI and the MOT further expanded their financial incentives. The EA introduced a new subsidy programme for local governments and for the construction of fuel stations. MITI introduced a new subsidy programme for electric vehicles and increased subsidies for vehicles powered by natural gas by 80 per cent. The MOT also expanded its subsidies for buses (*Enerugî to Kankyô*, 18 January 1996). Local governments and, with the MOT's guidance, the Japanese Truck Association started to collaborate with government policy. Combining the subsidy programmes of local authorities, MITI, the MOT and the Japanese Truck Association, members of the Truck Association could obtain up to a 100 per cent subsidy for the price differential between a conventional vehicle and one powered by natural gas.

While the EA, MITI and the MOT promoted LEVs, they (especially the EA and MITI) failed to coordinate their policies sufficiently. The primary policy objective of the EA was to reduce emissions of nitrogen oxides in urban centres. MITI, on the other hand, sought both to promote alternative energy sources and to maintain the competitiveness of the Japanese automobile industry in the face of the US initiatives. Indeed, MITI was reluctant to use the term 'low-emission vehicles', but preferred 'clean-energy vehicles' or 'alternative-fuel vehicles'. In accordance with their policy objectives, the EA and MITI established their own policy frameworks for LEV promotion, namely, the Automobile Nitrogen Oxide Law and the New Energy Law, and set separate targets for the introduction of LEVs (see below). There was no order agreed among the EA, MITI and the MOT for the priority to be afforded the different types of LEV. Their R&D projects were also conducted separately. In 1994 the EA's study group on LEVs concluded that, in the absence of ministerial coordination, the measures introduced were prone to inefficiency (EA, 1994c).

As Table 6.4 shows, the actual introduction of LEVs was far slower than these various targets would indicate; however, one exception was hybrid cars (see Figure 6.4). In 1997, Toyota marketed a hybrid car for the public and specifically promoted it in relation to the issue of global warming in the run-up to COP3. It received a good public response. MITI quickly introduced a new subsidy programme for the purchase of hybrid cars by individual consumers. Toyota's successful marketing of hybrid cars provoked competition among car manufacturers and, according to an EA

Table 6.4 *Targets for the introduction of LEVs and actual usage*

Programme	Target for the year 2000	Actual numbers in fiscal year 2000
Guideline for New Energy Introduction	490,000 (including LPG vehicles)	62,032 (excluding LPG vehicles)
Automobile Nitrogen Oxide Law	300,000 in designated urban areas in 6 prefectures	As above
Action Programme for the National Government to Take the Initiative in Pursuing Environmental Conservation as a Business Operator and Consumer	10% of government's vehicles	4% (from 0.1% in fiscal year 1997)

LPG: liquid petroleum gas.
Source: NEF (2001a); EA press release, 12 September 2001.

Figure 6.4 *Rates of introduction of LEVs, 1990–2003.*

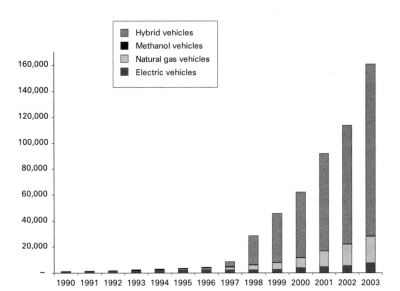

Source: EA, www.env.go.jp/doc/toukei/ (accessed 16 July 2005).

official, ended the general reluctance among car manufacturers to acknowledge that the technological capability for attaining a high level of fuel efficiency exists: they had feared that an open admission of such a technological capability might lead to more stringent fuel efficiency standards (interview with Kenji Fujita, 2000). It should be noted, however, that industry did not necessarily shift its focus from larger cars to LEVs. While promoting LEVs, the industry also responded to the general preferences of the public for larger (and therefore inefficient) cars (Ezawa, 1998: 106–11). Nevertheless, the new market climate was certainly not an adverse factor when the government, MITI and the MOT decided to introduce a 'top-runner approach' to the setting of fuel efficiency standards just before COP3 and the average fuel economy of petrol vehicles rapidly improved. Government policy thereby established a socio-economic milieu in which LEVs could be successfully promoted, but it was industry's long-term strategic view that actually achieved progress.

Traffic management and reduction:
institutions and interests hindering policy change
In addition to people's choice of larger cars, the growth in the distances travelled by car was another important factor contributing to the rapid increases in emissions of carbon dioxide from the transport sector. It is widely thought that this increase in car travel was accelerated by the liberalisation of oil imports in 1996, which reduced petroleum prices by up to 30 per cent in a short period (Kamei, 2002: 72–3) .

When the problem of global warming became a national issue, there was already a recognition within government that something must be done about the growth in road traffic. In addition to the worsening congestion and environmental problems of noise and air pollution, there also emerged a labour shortage in the road freight industry, as a new business practice of 'just-in-time' delivery prevailed. The EA, in a debate on nitrogen oxide policy, considered restrictions on vehicle use in polluted areas. The MOT also acknowledged the need to restrain traffic volume and stressed the importance of revitalising a declining public transport sector and promoting a meaningful shift in the types of freight transport used and a rationalisation of freight transport overall. The MOC, in formulating the Eleventh Five-Year Road Improvement Plan (1993–97), introduced the idea of transportation demand management (TDM), as a measure to supplement the existing policy of congestion relief based on an expansion of road capacity. The MOC's TDM was basically aimed at improving traffic flow through traffic dispersion and the efficient use of roads; it had little to say on constraining the volume of traffic.

Measures to reduce emissions of carbon dioxide by road traffic included the promotion of public transport, 'multimodal' transportation (i.e. the use of two or more modes of transportation based on an integrated transportation system), TDM, rationalisation of freight transport and road-building. Strategic promotion of intelligent transportation systems by the relevant ministries, as well as industry and academics, in collaboration since the mid-1990s, was also expected to be very much conducive to a reduction in emissions of carbon dioxide. However, the reduction objective in these policies was typically an afterthought. It was only in the run-up to COP3 that the transport-related ministries were brought together to discuss these policies in terms of reducing emissions

Figure 6.5 *Projected effects of new policies for reducing emissions of carbon dioxide in the transport sector, 1994–98.*

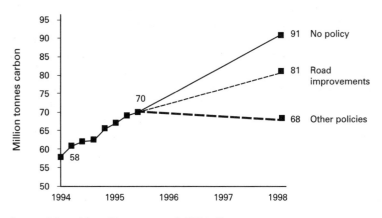

Source: Adapted from Kanemoto *et al.* (2003: 8).

of carbon dioxide. In September 1997, the MOT, MITI, the MOC, the National Police Agency and the Ministry of Post and Telecommunications together decided a carbon dioxide reduction programme in the transport sector. As Figure 6.5 shows, the government's policy was characterised by a lopsided reliance on road-building, in so far as other policies would have achieved a greater reduction in emissions.

The expansion of road capacity, as a major policy to alleviate traffic-related environmental problems, was implemented by the MOC and supported by industry, including Keidanren, the Japanese Truck Association and the Japan Automobile Manufacturers' Association. In the 1990s, however, this approach was increasingly criticised by environmental, citizens' and pollution victims' groups, which were concerned about the environmental limits to the absorption of pollutants and the effects of increases in the volume of traffic encouraged by new roads. The validity of road expansion as a measure to reduce emissions was also questioned by the UN FCCC review team which examined Japanese global warming policy in 1996 (Oh *et al.*, 1996). The MOC did not necessarily reject the idea of 'induced traffic'; however, its view was that, in Japan, the positive effects of new roads would outweigh the negative.

In the latter half of the 1990s, the government's trunk road policy came under increasingly severe attack. First, the need to build further, less cost-effective, expensive roads was questioned when the state deficit was mushrooming. Second, and more critically, in 1995 the Supreme Court found in favour of the plaintiff in a case concerning noise and other nuisances caused by traffic on National Route Number 43 and Hanshin Highway. Although the Court rejected the plaintiff's demand that traffic be restrained to a certain level, the first ever Supreme Court judgement admitting the liability of the state and the public road corporation in charge of the highway was a serious blow to the legitimacy of the MOC's road-building policy.

After the Supreme Court ruling the MOC convened its Road Council to consider policy on roads and the environment. Its interim report, issued in June 1997, urged the MOC to change the existing policy to place more emphasis on environmental protection and to enhance public participation in the policy process. Global warming policy was also discussed, and the Council called on the MOC to take action to contain traffic volume by integrating transport infrastructure policy and urban planning, and introducing stronger TDM measures, such as the use of economic instruments and traffic restrictions in particular areas. These views and policies were incorporated in the next (the Twelfth) Five-Year Road Plan (1998–2002). Moreover, in formulating that Road Plan, the MOC introduced two institutional innovations. One was an approach which incorporated the public's opinions on roads in the formulation policy. The other was the introduction of a manual on the cost–benefit appraisal of roads; this was made public, and the MOC indicated that such appraisals would include a 'no road' option.

These policy developments, however, soon proved to be minor policy adjustments in the face of social pressure against, and the physical limit to, the existing supply-side approach to traffic problems. The new Road Plan, after all, won an increased budget of ¥78 trillion for road-building when the budget for public works projects as a whole was reduced. Moreover, despite the fact that the MOC's Road Council called for a policy shift through the incorporation of environmental considerations into road policy in its interim report, in the end it revived the traditional approach of building more roads when it came to concrete measures to deal with the problem (Government of Japan, 1998).

There were two key reasons for this policy continuity. First, the method of cost–benefit analysis had a critical defect. The 'costs' considered in the method were only those for the construction and maintenance of roads. Neither the environmental costs nor those associated with the consequent increases in traffic induced by the new roads in the medium to long term were counted (Road Bureau, Ministry of Construction, 2003). Second, through the public involvement approach, it was shown that people wanted good roads. Though they demanded that environmental and social considerations be taken into account when individual new roads were built, few people were concerned about the broader transportation system and the environmental effects of more road-building. The apparent innovations, indeed, made real policy change more difficult.

More important in understanding the continuity in the government's policy of building more roads are, however, the institutions underpinning that policy. Road-building is implemented according to the Five-Year Road Plan decided by the Cabinet under the Emergency Measure Law for Road Improvement. The Road Plan is drawn up by the MOC, without Diet check on details, in accordance with a long-term Comprehensive National Development Plan. When the Fourth Comprehensive National Development Plan (1987 to around 2000) was established, the Cabinet decided a target of 14,000 km of new high-standard roads, and Road Plans were formulated to achieve this target. Moreover, when the Twelfth Five-Year Road Plan was being formulated (with the new the public involvement approach), the target of 14,000 km was quietly left to the next (the Fifth) Comprehensive National Development Plan (1998 to around 2010/15), to avoid critical scrutiny.

An equally important institution relating to road-building is the Earmarked Fund for Road Improvement, which is financed by taxes on the possession and use of road vehicles. More than 70 per cent of motoring-related taxes are used for roads (see Table 6.5). After the introduction of a specific taxation system, the MOC was able to raise revenue whenever it judged necessary to implement a five-year road plan. These earmarked revenues now finance more than 90 per cent of the central government's expenditure on roads, and account for nearly half of all road-building/improvement projects (Igarashi and Ogawa, 1997). This financial mechanism appears to have been very effective in extending road capacities

Table 6.5 *Motoring-related taxes*

Tax	National or local	Use of revenue	Share in motoring-related taxes
On purchase of vehicle			
Automobile acquisition tax[a]	Local	Prefectural and municipal governments' funds earmarked for roads	About 6%
On possession of vehicle			
Motor vehicle weight tax[a]	National	60% for national funds earmarked for roads; 15% for general budget; 25% for municipal governments' funds earmarked for roads	About 14%
Automobile tax	Local	Prefectural governments' general budget	About 22%
Light-vehicle tax	Local	Municipal governments' general budget	About 1.5%
On use of vehicle			
Gasoline tax[a]	National	National government's fund earmarked for roads	About 35%
Local road tax (tax on petrol)[a]	National	Local governments' funds earmarked for roads	About 3–4%
Diesel delivery tax (tax on diesel)[a]	Local	Prefectural governments' funds earmarked for roads	About 16–17%
Petroleum gas tax[a]	National	50% for national funds earmarked for roads; 50% for prefectural government's funds earmarked for roads	About 0.3%

[a] Earmarked for road improvement.
Note that the consumption tax that is applied upon the purchase of vehicles and fuels (presently set at 5 per cent) is not included here.

Figure 6.6 *Earmarked road expenditure and total length of roads, 1963–2001.*

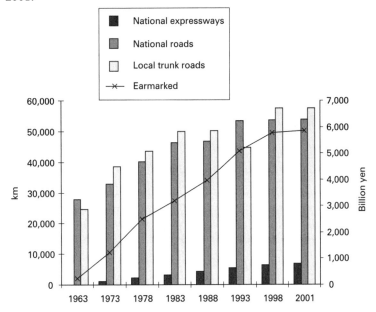

Source: Ministry of Land, Infrastructure and Transport (various years).

(see Figure 6.6) and gave rise to a growth spiral both in the number of cars and in the distance travelled by road vehicles (see Figures 6.7 and 6.8). As long as a stable source of finance for roads exists, road-building continues regardless of stagnation in public investment in the general budget.

Hand in hand with these institutions are interlocking interests in building roads among bureaucrats, politicians and the construction industry. Roads are an excellent 'pork barrel' for politicians, who depend on the vote-collecting capacity of the construction industry. There are also a number of road-related quasi-governmental and non-governmental bodies which offer lucrative posts for retired MOC officials. The promotion of road-building is politically and institutionally embedded in transport policy.

Furthermore, even if local governments wanted to formulate and implement an integrated policy for traffic reduction, they face

Figure 6.7 *Trends in passenger transport volume (million passenger-km) by type of transport, 1965–2001.*

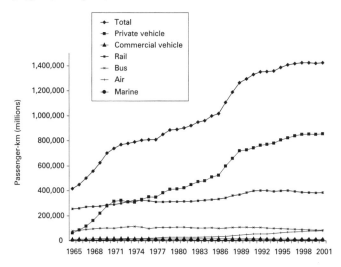

Source: Energy Conservation Centre (2003).

Figure 6.8 *Trends in freight transport volume (million tonne-km) by type of transport, 1965–2001.*

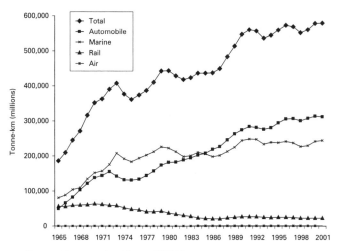

Source: Energy Conservation Centre (2003).

the institutional barriers of limited power and discretionary grants. At the same time, policy coordination at the central level is lacking not only between ministries but also between bureaus within the same ministry (Hayashi and Nagamine, 2001). Local governments are thereby bounded by compartmentalised transportation policy-making at the central level.

In reviewing progress on the integration of environmental consideration into transportation policy, the Central Environment Council concluded that while there were various measures conducive to environmental protection, environmental protection remained a peripheral objective. After all, there was still no measure aimed specifically at reducing traffic and no serious debate took place on the extent to which and how car dependency should be reduced (EA, 1998: 43). Government policy remained 'roads first'.

Conclusions

A number of both positive and negative corporatist traits can be detected in policy integration in Japan. Notable progress was made through policy concertation between the government and industry, especially in LEV policy. The government both steered industry to form a new industrial association, and collaborated with the haulage industry association to promote LEVs. Financial incentives were fully used. In the energy policy area, MITI and Keidanren reached a consensus on a non-zero-sum approach to energy, the economy and the environment. MITI was also pro-active in incorporating global warming concerns into its energy policy.

At the same time, however, consensus corporatism provided industry with good channels to influence government policy. In the case of new energy, pressure on various fronts to introduce a scheme requiring power companies to purchase electricity generated from these sources resulted in, perhaps, the second-best solution of the voluntary purchase scheme.

A more severe constraint on policy integration was the exclusive character of the Japanese corporatist institutions. In both the energy and the transport policy areas, environmental actors had little access to the locus of policy-making, and hence limited influence. One exception was a successful campaign against a new coal power plant, in which direct local protest worked, but this did not trigger any notable change to the existing politics or course of policy. The introduction of public involvement in road policy was a missed opportunity. As Mie Asaoka, the representative of the Kiko Forum, regretted, there was no provision for stimulating public debate or of transparent mechanisms necessary to reflect public voices in decisions (interview, 2000). It should be noted that the constraint of corporatist exclusion in Japan resulted not only in the failure of energy and transport policy-makers to incorporate environmental interests adequately into their policy-making institutions, but also in the poor political articulation of environmental interests. Asaoka said that 'the long LDP reign has built up a system that renders the public politically inert' (interview, 2000). Thirty-eight years of the producer-dominated corporatist political framework (under the LDP to 1993) has emasculated societal vitality, which is of great importance to the environmental capacity of a society. At the same time, the government did not take the lead in organising and integrating environmental interests in policy processes. This contrasted with its enthusiasm in organising economic interests, as seen in the case of LEV Forum.

One notable constraint on policy integration in Japan was strong past policy commitments, which were often institutionalised. In particular, the institutionalisation of the promotion of nuclear energy and road-building policy and tight policy networks associated with them left little political opportunity for effective challenges from the outside to the core policy networks, including parliamentary policy scrutiny.

The cases of nuclear energy and road-building policy also indicated the importance of interests. Here the issue-based approach, which borrows from public choice theory, gives insight. The close relations among the core political actors in energy and transport policy-making were often based on interlocking self-interest. Accordingly, these actors had a strong motivation to maintain the existing policy and relations by excluding potentially disruptive elements. Exclusive corporatist institutions and interests reinforced each other.

Interests also help to explain the initially positive reaction from energy policy-makers. More than 90 per cent of carbon dioxide emissions in Japan come from energy-related sources, hence emission reduction is largely a concern of energy policy. It was in

MITI's bureaucratic interests to respond quickly to the imperative of policy integration, so that its jurisdiction was defended and even expanded. More generally, once politicians had stressed the importance of global environmental issues, it was in the ministries' interests to talk them up in order to raise their share of the budget. Combined with the corporatist institutions of interventionism, jurisdictional competition gave policy-makers the incentive to take a (somewhat) proactive response to the integration imperative.

Thus, from one perspective, as an EA official said, the environment became an important element of policy-making in any sector, and government ministries were approaching the issue more positively and seriously than ever before (interview with Kenji Fujita, 2000). However, in so doing, MITI, in particular, effectively served its self-interests while keeping developments on policy integration under its control and its policy network closed. The global warming issue was, in this sense, co-opted rather than integrated.

This analysis may indicate that ministerial jurisdictional competition, combined with corporatist proactivism, had positive effects on policy integration. However, it also had serious constraints on policy integration, as was seen in the failure to coordinate LEV policy between MITI, the MOT and the EA. The *Report of the In-depth Review of the National Communication* on global warming policy in Japan criticised the lack of policy coordination: 'Agencies and ministries ... act independently in a spirit of competition rather than co-operating fully in [carbon dioxide] mitigation efforts' (Oh *et al.*, 1996: para. 13).

Moreover, a corollary of policy co-optation with the (systematic) exclusion of environmental interests is that the environmental concern is relegated to the status of a secondary issue, as the case of coal-related policy showed. Also, important conflicts were often ignored or obscured. The lack of progress or slow progress in tackling the relationships between electricity liberalisation, energy prices and energy use, between prices of cars and consumer preferences for larger vehicles, and between road-building, traffic volumes, congestion and emission levels are cases in point.

Looked at in this way, policy integration in Japan was constrained as much as driven by corporatist institutional characteristics, or more accurately by the interactions between institutions and interests. Institutions and interests reinforced each other. As the Central Environment Council concluded, environmental considerations were not adequately integrated into transport policy (EA, 1998), and the integration of global warming concerns into energy policy was largely at the level of rhetoric, with ambitious targets not substantiated by commensurate policies and measures.

Notes

1   Peak demand for electricity comes during the daytime in the summer, when people extensively use air conditioning. During the 1980s, the peak demand for electricity increased by the equivalent of more than two nuclear power plants' capacity every year (*Asahi Shimbun*, 9 November 1991).
2   Power companies would have been obliged to buy only 'surplus' electricity from non-power companies that generated power for their own use (e.g., private power generation from photovoltaic panels, public authorities generating energy from waste and factories generating power for their own use). They would not have to accept electricity which others generated to sell on.
3   With the administrative reorganisation in 2001, the Science and Technology Agency was incorporated within the Ministry of Education, Culture, Sports, Science and Technology.
4   In 2003 MITI decided to introduce a low-level tax on coal. The use of coal as a raw material is exempted.

7

Policy developments in Britain on global warming: in search of political leadership

Britain accounts for only about 2.5 per cent of total world emissions of carbon dioxide. In the early stages of global warming politics, this was often used by the government as an excuse not to take 'hasty action'. Such cautious attitudes were reinforced by the government's preoccupation with the biggest political project of the day, privatisation of the electricity industry. Britain was involved in the FCCC and its negotiation processes both as a member of the EU and as an independent state.

About 80 per cent of British GHG emissions are in the form of carbon dioxide, of which about 97 per cent is energy-related. As Table 7.1 shows, overall carbon dioxide emissions fell steadily from 1970. Emissions from industry and, to a lesser extent, households declined, while those from the transport sector rapidly rose. However, as seen in Figure 7.1, the downward trend started to lose its pace since the mid-1990s. After 1997, carbon dioxide emissions from the transport sector consistently exceeded those from the domestic sector and in 2002 they just exceed those from the industrial sector (Figure 7.2). British emissions of carbon dioxide have been largely determined by the use of coal and their long-term decline is basically explained by the rapid decline in the use of coal (see Figure 7.3). The shift in economic structure from traditional heavy manufacturing industry to light manufacturing and services, which now account for about 70 per cent of total output (in terms of GDP), is also an important factor contributing to the downward trend in emissions.

Table 7.1 Trends in carbon dioxide emissions (millions of tonnes of carbon) by end users, 1970–2002

	1970	1980	1990	1996	2000	2002	Percentage change, 1970–90	Percentage change, 1990–2002
Industry	82.7	63.1	50.8	44.9	43.5	40.8	–38.6	–19.7
Transport	20.7	25.9	39.2	41.2	41.7	41.5	+89.4	+5.9
Domestic	55.3	49.4	41.8	41.8	41.5	40.5	–24.4	–3.1
Commercial and public administration	23.3	23.0	23.2	21.6	21.1	19.7	–4.3	–15.1
Others	3.4	2.9	4.4	5.3	3.5	3.8	11.7	–13.6
Total	185.3	164.3	159.3	154.8	151.3	146.3	–14.0	–8.2

Source: DEFRA (2004a).

Figure 7.1 *Carbon dioxide emissions by source, 1970–2003.*

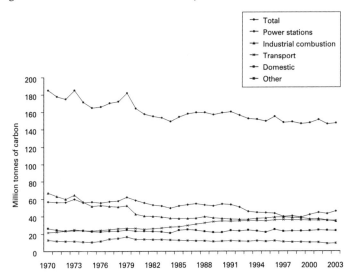

Source: DEFRA (2004a).

Figure 7.2 *Carbon dioxide emissions by type of end user, 1990–2002.*

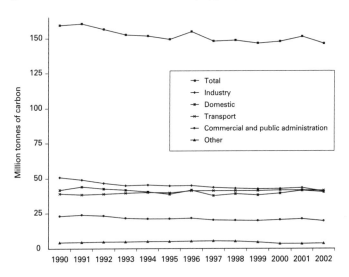

Source: DEFRA (2004a).

Figure 7.3 *Energy supply by fuel type, 1970–2002.*

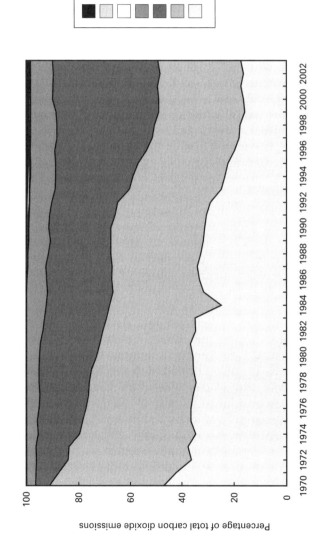

Renewables and waste
Net electricity import
Hydro-electricity
Nuclear electricity
Natural gas
Petroleum
Coal

Percentage of total carbon dioxide emissions

1970 1972 1974 1976 1978 1980 1982 1984 1986 1988 1990 1992 1994 1996 1998 2000 2002

Source: DTI (2004).

The emergence of global warming on the British political agenda

The global warming issue abruptly rose up the British political agenda after the sudden interest in environmental issues, particularly those of a global dimension, by the then Prime Minister, Margaret Thatcher. In September 1988, in her speech to the Royal Society, she said that 'Protecting ... [the] balance of nature is ... one of the great challenges of the late twentieth century'. Moreover, at the Conservative Party conference the following month, she stated that 'we Conservatives ... are not merely friends of the Earth – we are its guardians and trustees for generations to come.... All we have is a life tenancy – with a full repairing lease. And this government intends to meet the terms of that lease in full' (Harris, 1997: 326–46). This was an about-face (at least on the surface) in her stance on the environment, which she once called a 'humdrum' issue.

Her speeches were followed by immediate developments in British environmental politics. In July 1989, Nicholas Ridley was replaced as Environment Secretary by Chris Patten, who, in contrast to Ridley, was well known for his sympathetic attitude towards the environment. The first white paper on the environment was also promised. Political interest specifically in global warming also took off. Just one day after her speech to the Royal Society, the Foreign Secretary called on the UN General Assembly to debate the issue of climate change (Boehmer-Christiansen and Skea, 1991: 264). In January 1989, Thatcher chaired a Cabinet meeting attended by ministers from eight government departments, and this was followed by a day-long seminar on climate change to which six ministers and about thirty scientists and industrialists were invited (*The Times*, 12 January 1989; *The Economist*, 8 April 1989: 34, 38). Also, building on Britain's world-leading climate science, the Hadley Centre for Climate Prediction and Research was established as part of the Meteorological Office, with DOE funding. The Centre became the operating agency for the IPCC's scientific Working Group. In the environment white paper published in September 1990, global warming was posited as 'one of the biggest environmental challenges now facing the world' (Her Majesty's Government, 1990: para. 5.1).

There are various speculations on Thatcher's green conversion. Some argue that she wanted to destroy the political power of the mineworkers and to promote nuclear energy. In December 1988, in the House of Commons, she said that the greenhouse effect 'is only partly a matter of energy efficiency; it is also a matter of preserving tropical rain forests and of replacing some coal energy production by nuclear energy' (*Hansard*, 15 December 1988, col. 1080). Others believe it reflected Thatcher's desire to take the leadership in international politics (McCormick, 1991: 65). The global environment was rapidly attracting the interest of world political leaders, perhaps as the next major issue in a new era of détente (Yonemoto, 1994: 52–60).

Against these sceptical views, however, David Pearce, a former part-time adviser to the Environment Secretary, says that Thatcher, who was a scientist, genuinely believed in global warming and that she always emphasised 'good science' (interview, 1998). Thatcher was also directly influenced by some individuals close to her, including the former British diplomat Sir Crispin Tickell (who had taken a major interest in the science of climate change since the mid-1970s), senior scientists and members of the Royal Society. In particular, between 1987 and 1990, scientists associated with the Meteorological Office and the DOE had several opportunities to brief her (Boehmer-Christiansen, 1995: 177).

Thatcher's green conversion, whether it was genuine or not, soon raised broad suspicion and criticism, as she repeatedly emphasised the importance of nuclear energy while discounting the importance of energy efficiency. Moreover, Thatcher appeared to begin to lose her initial enthusiasm when it became clear that tackling global warming had significant implications for freedom of choice, non-intervention, free markets and other issues that she had strongly promoted (McCormick, 1991: 66–7). Nevertheless, Thatcher's speeches were milestones. By the time she resigned the premiership in November 1990, environmental issues had been 'legitimised', to be addressed positively, and the game of environmental politics had changed (see McCormick, 1991: 61).

Developments in setting a target

Despite the early political developments initiated by Thatcher's green conversion, the government was very cautious about making specific commitments when the EU tried to take the lead in the emerging international politics of global warming. At the

Noordwijk conference in November 1989, when the EU was pushing Japan, the USA and the former Soviet Union to agree to a specific carbon dioxide reduction target, Britain sided with these reluctant countries, and stressed the scientific uncertainties. Nevertheless, Britain, as an EU member state, agreed to the Noordwijk Declaration, which stated that carbon dioxide emissions should be stabilised no later than 2000. Thereafter, Britain took three steps on the target issue during the Conservative era: in May 1990, Thatcher announced the first British target; in March 1992, the target was moved forward to meet the *de facto* international norm; and in March 1995, the then Environment Secretary, John Gummer, called for 'challenging but achievable' targets towards 2010 for developed countries.

The first British target: May 1990
In May 1990, a meeting of the UN Economic Commission on Europe was held in Bergen, Norway (the Bergen conference). At the end of the conference Britain, for the first time, accepted the need to set a national carbon dioxide limitation target (*The Times*, 18 May 1990). This policy change followed the release of an interim report, just before the conference, from the IPCC's WG-I, which assessed climate science and was chaired by a leading British scientist. The report was accepted almost without qualification by Thatcher. Ten days after the conference, she announced that Britain would adopt as a national target stabilisation of carbon dioxide emissions at the 1990 level by 2005. Moreover, she said that 'provided others are ready to take their full share, Britain is prepared to set itself the very demanding target of a reduction of up to 30 per cent in presently projected levels of carbon dioxide emissions by the year 2000'. This more or less amounted to a reduction to the 1990 level by 2000 (*ENDS Report* 184, 1990: 11).

Although this was a watershed, it was not free from criticism. The absence of similar commitments from economic competitors like the USA and Japan meant that the British commitment was stabilisation by 2005, but this was not consistent with the Noordwijk Declaration, which demanded stabilisation by 2000. Within a week of Thatcher's announcement, the then European Environment Commissioner, Carlo Ripa di Meana, publicly appealed to Thatcher to change the target year to 2000 (*ENDS Report* 184, 1990: 12). Britain was resistant. There was an attempt by Environment Secretary Chris Patten to bring the target forward to 2000 in the preparation of the September 1990 white paper but this was blocked by strong opposition from other ministries. The 2000 target was simply thought too costly (Madison and Pearce, 1995: 127).

For those who made the case for 2005, even this was considered 'fairly challenging' within the context of the emission projections at that time. Although there was ample potential for reducing carbon dioxide emissions, there were considerable uncertainties about the future of nuclear power stations and the extent to which privatisation of the electricity industry would stimulate the switch from carbon-intensive coal to less carbon-intensive and cheaper gas. Moreover, a previous energy efficiency campaign run by the government had failed to achieve the desired effects (Madison and Pearce, 1995: 126). The government's stance was, however, attacked by environmental activists. The *ENDS Report*, for example, drawing on calculations made by FOE and the ACE, pointed out that there had been negligible changes in Britain's carbon dioxide emissions since 1985, as well as the real potential for further improvements in energy efficiency on cost-effectiveness grounds alone (*ENDS Report* 184, 1990: 13–14). Moreover, Thatcher's target was not followed by an elaboration of the specific measures needed to achieve it, making it rather a political gesture.

At the joint EU Energy and Environment Council in October 1990, Britain agreed to the target of the EU as a whole to stabilise carbon dioxide emissions at the 1990 level by 2000. But it did not change its own 2005 target, isolating itself from other major EU member states (*ENDS Report* 190, 1990: 11–13).

Stabilisation by 2000: March 1992
In March 1992 the then Environment Secretary, Michael Heseltine, announced that Britain was prepared to speed up its timetable provided that other countries did likewise (*Financial Times*, 3 March 1992). The announcement came a while after Energy Paper 59 (EP 59) was released, in December 1991, which decreased the official projections of Britain's future carbon dioxide emissions (DTI, 1992). As Table 7.2 shows, the new projections indicated that the 2000 target was achievable with less pain and was preferable to the 2005 target.

Table 7.2 *Projections of carbon dioxide emissions in Energy Paper 55 (1989) and Energy Paper 59 (1992) (million tonnes of carbon)*

	Old projections (EP 55)	New projections (EP 59)
1985	158	–
1990	–	160
2000	174–206	156–178 (170)
2005	178–225 (212)	166–200 (183)
2020	188–316 (234)	188–284 (221)

Figures in parentheses are the projections based on the assumption of central economic growth with low energy prices.
Source: DTI (1992: 8); *ENDS Report* (203, 1991: 20).

However, EP 59 did not bring an immediate change in government policy. The then Energy Secretary, John Wakeham, was reluctant to bring the target forward. His view remained that the 2005 target was demanding enough and that there was still considerable uncertainty about future carbon dioxide emission trends (*Hansard*, 6 December 1991, cols 258–9w). In fact, EP 59 cautiously concluded that the trend in carbon dioxide emissions since the mid-1980s had been slightly upwards if temperature effects were allowed for (DTI, 1992: 35–6). Nevertheless, in the run-up to the Rio Summit, Prime Minister John Major, supported by Heseltine, took the leadership in bringing the target forward. It is said that the government did not want to be left behind other OECD countries, since Major was the first political leader from among those countries to announce his intention to attend the Rio Summit in June 1992. By this time most other OECD states had a target for 2000 (Madison and Pearce, 1995: 127–8). The target was, thus, brought forward, but with the condition that other developed countries took similar action.

The cautious attitude continued even after Britain signed the FCCC at the Rio Summit. The government did not want to go beyond obligations set out by the Convention, which had largely been written by the British and US delegates. At the EU Environment Council in December 1992 Michael Howard, as Environment Secretary, refused to reaffirm a declaration made by EU environment and energy ministers in October 1990 that the EU was committed to stabilising its carbon dioxide emissions at the 1990 level by 2000. The reasons were twofold. First, the text of the FCCC did not say 'stabilisation' but only 'return', with no implication for carbon dioxide emissions beyond 2000. British resistance to stabilisation is understandable, given the projections of rapid increases in carbon dioxide emissions after 2000. Second, the EU commitment was conditional on other industrialised countries making similar commitments, so until the USA pledged itself to a similar target there was no EU commitment (*ENDS Report* 216, 1993: 37). In March 1993 at the EU Environment Council, Britain again maintained that it was 'far too early' to sign up for stabilisation. In the end, however, British resistance was suppressed and the EU's 'stabilisation' objective was clarified.

Britain ratified the FCCC in December 1993. It was the first EU member state to do so when the EU member states were still in dispute with each other over the introduction of an EU carbon/energy tax as a condition for the ratification of the EU as a whole. Britain, in a sense, bypassed the EU framework. It was symbolic of the assertion of British sovereignty *vis-à-vis* the EU.

The 'challenging' target: March 1995

In the run-up to COP1 in spring 1995, it became clear that Britain was one of only a few countries likely to meet the 2000 target. Most developed countries were reluctant to go onto the next phase (namely, commitments beyond 2000). Even the EU proposed only that carbon dioxide emissions should not be increased after 2000 or, in other words, simply stabilisation of emissions. In such a context, Britain jumped into the position of being a world leader. When the developed countries narrowly agreed to initiate a process to strengthen commitments at COP1, the then Environment Secretary, John Gummer, called on developed countries to reduce all GHGs to 5–10 per cent below the 1990 levels by 2010.

This new British policy was adopted after radical revisions of the carbon dioxide emission projections. Energy Paper 65 (EP 65), which was completed just before COP1, forecast that Britain's carbon dioxide emissions would fall by 4–8 per cent from the 1990 level by 2000 if account was taken of the existing measures against global warming, and then rise to just over the 1990 level by 2010 (DTI, 1995a). However, other GHGs were expected to fall considerably; as Energy Minister Tim Eggar stated, a 5–10 per cent cut in total GHGs 'can be done through a variety of measures that are already in place' (*ENDS Report* 242, 1995: 24). The new projection no doubt led the government to change its previous gas-by-gas approach to a basket approach which treated GHGs collectively, and Gummer was able to win Cabinet approval of the apparently ambitious target on the grounds that Britain would have to do little to achieve it (*ENDS Report* 242, 1995: 24).

In March 1997, at the EU Environment Council, the British government agreed to a 10 per cent cut in GHG emissions (carbon dioxide, methane and nitrous oxide) to contribute towards the EU's proposed target for COP3 negotiation of a 15 per cent cut in total EU GHG emissions, to 1990 levels by 2010. In doing so, Britain accepted the 'burden-sharing' approach to which it had been strongly opposed.[1] Britain, however, refused to go further than a 10 per cent cut. It also objected the establishment of a separate reduction target for carbon dioxide. British carbon dioxide emissions were projected to increase after 2000 from the 1990 level (*ENDS Report* 266, 1997).

National strategy

The first formal proposals on measures to reduce emissions of carbon dioxide were made in the 1990 white paper. There were, however, few new measures – most were measures that had already been taken for other reasons. The white paper gave no indication of plans to introduce energy taxation or other measures to raise energy prices until the next general election, but signalled possible increases in road fuel duties (Her Majesty's Government, 1990). The government was warned by the House of Commons Energy Committee not to rely too much on higher energy prices or impose costs on the economy in pursuit of a reduction in carbon dioxide emissions before 'energy efficiency measures with net benefit to the economy are being fully exploited' (House of Commons Energy Select Committee, 1991: para. 150).

After Britain signed the FCCC the government started to formulate a national strategy to cut carbon dioxide emissions, as required by both the Convention and the EU. The goal was to save 10 million tonnes of carbon from the projected 2000 level in the business-as-usual scenario. In December 1992, the DOE published a discussion document, *Climate Change: Our National Programme for CO_2 Emissions*, for consultation. In March 1993, the DOE held a series of workshops to discuss the document; these were attended by a wide range of interested parties, including public bodies, economic, environmental and social organisations and academics. The outcome of these workshops was reported and discussed at a major conference in May. More than 100 interested parties and most government departments participated in the conference. The government placed great importance on the *process* – simply having a broad and intensive consultation exercise – partly because it believed this would gain voluntary commitment to action from the participants. This British approach was well described by John Gummer:

> [there] is no way of delivering what we are concerned to deliver unless we change fundamentally the lifestyles of the whole of the population, and such a change cannot be achieved, even if it were sensible to try to achieve it, by diktat or by regulation ... an industry-driven system ... will be the best and most cost-effective solution. (House of Lords Select Committee on Sustainable Development, 1994: Q. 1871)

In July 1993 a government programme of action had been drawn up in time for the EU deadline for its submission and

Table 7.3 *The carbon dioxide saving plan set out in the 1994* UK Programme

	Carbon dioxide savings in 2000 (millions of tonnes of carbon)
VAT on domestic energy	1.5
Energy Saving Trust	2.5
Energy efficiency information and advice	1.5
Increased CHP target	1.0
Increased renewables target	0.5
Building regulations	0.25
Appliance labelling and minimum standards	0.5
Public sector energy efficiency	1.0
Increases in road fuel duty	2.5
Total	11.25
Total allowing for overlaps	10.0

VAT: value added tax; CHP: combined heat and power.
Source: Her Majesty's Government (1994a).

published in January 1994 in *Climate Change: The UK Programme* (Her Majesty's Government, 1994a). *The UK Programme* was praised by the House of Lords Select Committee on Sustainable Development (1994) as 'the most detailed response by the government to [a] major environmental concern'. It set forth (in para. 3.76) several interrelated, key principles to be followed in formulating and implementing policies and measures: first, carbon dioxide savings should be achieved in the most cost-effective way; second, government intervention should be avoided as far as possible; and third, where government action was needed, economic instruments were preferred to regulation. The policy was justified on the grounds of the precautionary principle as well as in terms of economic efficiency. Table 7.3 shows details of the programme.

VAT on domestic fuel and power
Value added tax (VAT) on domestic fuel and power was announced by the then Chancellor of the Exchequer, Norman Lamont, as a 'green' measure in his March budget in 1993. His intention was to introduce it from April 1994 at a rate of 8 per cent and at a full rate of 17.5 per cent from April 1995. The rationale was that all other EU member states had such a tax and that it would rectify the market distortion in which energy-saving products were taxed at the full rate while energy use was zero rated. While this was a 'green' tax, it was widely regarded as a measure to tackle the problem of the massive public deficit which the government created on 'Black Wednesday' in 1992. It was estimated that the VAT would bring £3 billion a year from 1995 onwards; if implemented fully this would have been the biggest revenue raiser in the budget and would have gone a long way towards solving the deficit problem (Lamont, 1999: 349). It should be noted, however, that although raising revenue was an important motivation, the environmental argument was also, according to Lamont, 'genuine'. He believed that to meet the carbon dioxide stabilisation target it was necessary to encourage greater energy conservation, and the most reliable means to achieve this was through the price mechanism (Lamont, 1999: 347–9). It was also said that when the government was strongly resistant to the EU's attempt to introduce an EU-wide energy/carbon tax, the government needed to show that Britain did not need such a measure to achieve its national target. The extension of VAT was, together with the fuel duty

escalator described below, a way of avoiding an energy/carbon tax, with exemption for industry.[2]

The proposal, however, provoked strong public resentment. It broke an electoral pledge not to raise taxes and, more importantly, touched on the long-standing problem of fuel poverty.[3] It is estimated that there were at least 5.8 million fuel-poor households in England in 1991 (DTI, 2000a: para. 4A.2–3). According to Lamont, however, the real hardship for the poor was likely to be relatively modest, as energy prices had been falling since the 1980s (Lamont, 1999: 347–9). To pacify public anger, Lamont promised a compensation package for vulnerable groups. In November 1994, the then Chancellor, Kenneth Clarke, outlined the compensation scheme in his autumn budget. The whole relief package was to cost £1.3 billion in 1996–97 compared with the £2.3 billion in revenue from the extension of VAT to fuel. One notable point here was that the revenue raised was quasi-earmarked for the relief package, which included subsidies for home insulation, in the form of the Home Energy Efficiency Scheme (HEES). It was an innovation because it broke the Treasury's traditional opposition to hypothecation.

The compensation scheme, however, did not allay public resentment. Several Conservative backbenchers, who anticipated a negative electoral effect from the new tax, joined Labour and the Liberal Democrats to pressurise the Chancellor into dropping the introduction of the full VAT rate. In December 1994, the government was defeated over the planned increase of VAT on fuel to the full rate (*The Times*, 7 December 1994, 28 November 1994; *Financial Times*, 7 December 1994) and therefore failed to fulfil the rationale of rectifying the market distortion. During the parliamentary debate the government made little reference to the imperative of carbon dioxide reduction.

Increases in fuel duties

Transport was singled out as the most important sector in tackling the carbon dioxide problem, as it was projected to be the main contributor to the increase in emissions. The government also recognised the limited effectiveness of information and advice in this area (DOE, 1992). The government's intention to raise fuel duties had been revealed in the 1990 environment white paper. However, what was innovative in Lamont's announcement in his 1993 budget was that he 'statutorily bound the hands of his successors' (O'Riordan and Rowbotham, 1996: 248). Fuel duties were to increase by an initial 10 per cent, with subsequent annual increases of at least 3 per cent, in real terms, scheduled to be implemented. In November 1993 Chancellor Kenneth Clarke raised this annual rate from 3 per cent to 5 per cent. By 1997 fuel duties had risen by an average of 6.75 per cent for unleaded petrol, 7.75 per cent for leaded petrol and 8.25 per cent for diesel per year in real terms (Her Majesty's Government, 1997b: para. 2.37). These increases made British fuel prices the highest in the EU. In response to the resultant growing concern about the effect of these rises on industry's competitiveness, the British government began to press for 'significant' increases in minimum road fuel duty in other member states.

There was doubt, however, about the effectiveness of the tax. As the government argued, in theory the expectation of the long-term increases in fuel duties gave consumers and manufacturers an incentive to respond. However, the scheme was not publicised well enough. Moreover, the continuing downward trend in pre-tax fuel prices at that time almost cancelled out the increases in fuel duties. For example, despite a 21 per cent increase in duty on unleaded petrol between 1994 and 1996, the real increase in price was only 1.5 per cent (RCEP, 1997: para. 2.44). As in the case of VAT, it was widely believed that the motivation was primarily a financial one painted green.

The Energy Saving Trust
The Energy Saving Trust (EST) was established as an independent body in November 1992 by the DOE, the Scottish Office, British Gas, fourteen regional electricity companies (RECs) in England and Wales and two Scottish electricity companies. The EST was a project funded by the Private Finance Initiative. The EST's main goal was an overall reduction in total energy consumption through the promotion of energy efficiency and, thus, to lessen the environmental impact of energy use. In the short term, as requested by the government, the EST prioritised the domestic and small-business sectors, which were thought to be particularly affected by the barriers to energy efficiency investment. The EST developed an 'integrated package' for the promotion of energy efficiency, from traditional measures such as the provision of information,

to a new initiative of financial assistance for consumers and a more innovative approach aimed at establishing energy efficiency markets. In the long term, it aimed to expand its work into the transport and industrial sectors. Within two years, the EST had achieved a number of successes.

The EST, however, soon faced a financial crisis. The government originally intended to finance it until the gas and electricity markets were fully liberalised in 1998,[4] with contributions from British Gas and the RECs, which in turn would impose a levy on gas and electricity bills in the form of an 'E' factor. The 'E' factor was a brainchild of the first Director-General of Ofgas, the regulator for the gas industry, Sir James McKinnon. The levy would be passed onto consumers only when the Director-General judged that the energy efficiency programmes proposed by British Gas were in the interests of consumers. The EST's budget for the first year was only £6 million, compared with the total £2 billion needed to achieve the target of reducing carbon emissions by 2.5 million tonnes.

The financial problems of the EST became even more serious in January 1994, when the new Director-General of Ofgas, Clare Spottiswoode, objected to the 'E' factor as a means of funding the EST, arguing that it was a 'regressive' tax which she had no statutory authority to impose, and that energy efficiency would best be promoted through competition (*ENDS Report* 229, 1994: 34). Meanwhile, although the electricity regulator approved a small levy on all households to finance energy efficiency measures, the money raised was not for the EST's projects: it was for the Energy Efficiency Standards of Performance scheme, as it was called, which was run by the RECs and overseen by the regulator, and the EST was given only a supporting role.[5] Moreover, the political developments associated with the introduction of VAT on domestic energy politically sensitised any measure which would increase domestic gas and electricity prices. With this financial turmoil, the EST was compelled to scrap some of its successful schemes.

The immediate cause of the problem lay in the institutional arrangements for the regulator. Both gas and electricity regulators were independent authorities, and their Directors-General were given discretionary power to fulfil their respective tasks imposed by the Gas Act and the Electricity Act. The government, for its part, was preoccupied with the creation of free markets and did not want to intervene in their activities. As a result, politically sensitive decisions on how to strike the balance between environmental and market competition objectives, and on what was in the best interests of the public, were left to politically non-accountable individuals.

Hoping to alleviate its financial crisis, the EST asked the government to incorporate in the Gas Competition Bill, due in February 1995, a mandatory energy efficiency levy on gas consumers and longer-term domestic gas contracts, so that its energy efficiency measures could produce a tangible benefit. Meanwhile, the House of Commons Trade and Industry Select Committee urged the government to take responsibility for the energy efficiency levy rather than leaving it to independent regulators. The EST, however, did not see this as a good solution, since if the government collected the levy the money raised would go to the Treasury, which does not, in principle, accept tax hypothecation. In the end, neither the request of the EST nor that of the Trade and Industry Select Committee was accepted by the government, which anticipated further bitter criticism in response to increases in energy prices following the unsuccessful attempt to raise VAT on domestic energy to the full rate (O'Riordan and Rowbotham, 1996: 252). The final version of the Bill made it the new, modest duty of the Director-General of Ofgas to take into account the effect on the environment of gas business activities (DOE, 1995: 4). While modest, this was a step forward, as the gas regulator had previously had no responsibility for the environment.

Early in 1996, the EST received a 'final blow' when British Gas withdrew from the 'E' factor funding scheme, which had been approved by Ofgas after a five-month examination, arguing that the scheme would not be a 'sensible business decision' (*ENDS Report* 253, 1996: 7). This raised the fundamental question of how far energy efficiency could be promoted in future, especially after the full liberalisation of the energy markets. In March 1995, the DOE took legal power to part-fund the EST to bear the majority of the EST's operating costs. In May the Environment Secretary, John Gummer, announced additional funding of £25 million a year until the energy markets were fully liberalised. This was the fruit of a shrewd move taken by Gummer to maintain the EST's operation as a core instrument for energy efficiency and reducing emissions of carbon dioxide. With the announcement by the Trade and Industry Secretary of privatisation of the nuclear industry in 1996, the levy on electricity bills to subsidise the nuclear industry was to be abolished at least eighteen months earlier than had originally been scheduled. The effect was to reduce electricity bills by about 8 per cent and this was expected to cancel out the carbon dioxide savings from VAT on domestic energy, resulting in increases in emissions. Gummer brought up the EST's case as a way to tackle this problem (DOE, news release, 11 May 1995; *ENDS Report* 244, 1995: 13).

Encouraging voluntary action

As the government believed that its leadership should be exerted in the field of the provision of authoritative information and advice (DOE, 1993), the encouragement of voluntary action was an important strategy. The central locus in carrying out this role was the Energy Efficiency Office (EEO) in the DEn and, after the abolition of the DEn in 1992, in the DOE.[6] The main activity of the EEO was the Energy Efficiency Best Practice Programme, launched in 1989 (Her Majesty's Government, 1994a: para. 2.57; DOE, 1995: 31). In 1991, partly to facilitate the Best Practice Programme, the then Energy Secretary, John Wakeham, also launched the 'Making a Corporate Commitment' campaign, which specifically prompted top management to sign a declaration of commitment to responsible energy management. The EEO also offered limited financial assistance to encourage those who could least afford energy efficiency measures to take action. In addition, the HEES provided low-income householders with advice and grants for basic energy efficiency measures. After its launch in 1991 the budget for the HEES steadily increased from £27 million in the first year to £75 million in 1997/98.

Initiatives were also taken to promote voluntary action orchestrated by business associations. The DOE, for example, carried out more targeted campaigns in cooperation with sectoral associations, such as the Building Employers' Confederation, the British Iron and Steel Producers' Association, the Heating and Ventilating Contractors' Association and the Electrical Contractors' Association. However, the role of trade associations in inducing voluntary action hardly went beyond the provision of information and advice, and some attempts to do so bore little fruit. The CBI established its Environment Business Forum, which offered guidance to its members, who in turn could commit themselves to an Agenda for Voluntary Action, which featured a requirement for environmental reporting. This attracted only around 200 businesses and failed to fulfil its potential. Also, the Advisory Committee on Business and the Environment tried to obtain voluntary energy-saving targets from six industries that were major emitters of carbon dioxide (iron and steel, chemicals, food, drink, tobacco and minerals) but this failed after an extended approach. The CBI's comments on the failure of the Advisory Committee are worth noting. According to them, the fact that in Britain fewer companies belong to trade associations made an effective voluntary agreement difficult. The CBI also raised the free-rider problem and doubted the validity of the approach (*ENDS Report* 244, 1995: 3). An implication was that the institutional capacity for negotiated voluntary agreements was lacking.

Nevertheless, the government looked for opportunities for voluntary agreements by those trade associations which were well organised and represented their industry well, expecting knock-on effects on the economy as a whole (interview with Terry Carrington, 1998). The prime target for voluntary agreements over energy efficiency was the Chemical Industry Association. The Association was a member of the industry's European federation, which launched a voluntary energy efficiency programme in 1992 with a target of improving energy efficiency in place of an EU carbon/energy tax. In 1997, the Chemical Industry Association and the Department of the Environment, Transport and the Regions under the incoming Labour government successfully concluded a voluntary agreement on an energy efficiency target, which was set at the same level as the voluntary target of the European chemical industry association. Developments at the European level paved the way for a new initiative to be realised in British environmental policy.

Regulations and labelling schemes
Although the government was committed to non-intervention, regulations were used to provide information to consumers in the form of energy labelling in combination with energy efficiency standards. A voluntary, and later mandatory, energy labelling scheme was introduced in building regulations, which contained insulation standards, and this was praised by the House

of Commons Energy Select Committee (1991: para. 95) as 'one of the most important recent developments in energy efficiency'. The government was also in favour of labelling, supported by minimum standards for appliances at the EU level. The voluntary approach was preferred, but the government was ready for mandatory labelling and minimum standards, where necessary (Her Majesty's Government, 1990: para. C.33). British manufacturers of appliances had already provided information on energy efficiency in the form of promotional literature. The government thought that to redisplay the information in a label form would not impose unacceptable additional costs on manufacturers and that the labelling would be more effective in providing information for consumers (Her Majesty's Government, 1990: para. C.31). The government was, however, cautious when it came to specifying minimum standards. Some appliance manufactures strongly opposed the EU initiatives, and governments of these manufacturers' countries, including Britain, were convinced of the need to defend their competitiveness.

Supply-side measures: targets for combined heat and power and renewable energy
Targets for the installation of renewable energy capacity and combined heat and power (CHP) facilities[7] were set out in the 1990 environment white paper, but in 1993, to achieve the remaining one-third of the carbon dioxide reduction requirement, they were strengthened. In the *UK Programme*, the CHP target towards 2000 was 5,000 MW and the renewables target 1,500 MW.

The principal instrument to promote renewable energy was the Non-Fossil Fuel Obligation (NFFO). The NFFO operated through the Electricity Act 1989, which empowered the Trade and Industry Secretary to make orders requiring electricity supply companies to secure specified amounts of non-fossil fuel energy by signing contracts with generators at premium prices, thus providing the generators with a guaranteed market. There were different tranches for nuclear and renewables. For the latter, contracts were granted to those generators that tendered the lowest prices. The additional costs for the electricity supply companies were covered by a fossil fuel levy on electricity bills. The levy was set at 11 per cent in 1991, but had fallen to 0.7 per cent by January 1997. To facilitate and simplify the dealing, the Non-Fossil Purchasing

Table 7.4 *Progress in the NFFO scheme as at 31 December 2000*

	Number of projects	Contracted capacity (MW)	Commissioned capacity (MW)
NFFO-1 (1990)	75	152.1	139.7
NFFO-2 (late 1991)	122	472.2	172.4
NFFO-3 (1995)	141	626.9	293.5
NFFO-4 (1997)	195	842.7	156.7
NFFO-5 (1998)	261	1,177.2	55.6
Total	794	3,271.1	877.9

Note: These figures are those for England and Wales only.
Source: DTI (2001).

Agency was established. The Agency signed the contract on behalf of the RECs and reimbursed them the costs of the premium price. Despite the programme's initial promise, by the latter half of the 1990s it had become increasingly clear that the target was going to be very difficult to achieve, essentially because of the problem of securing planning permission and finance (see Table 7.4). This problem will be explored in more detail in Chapter 8.

The CHP target was set to double the existing capacity. Although the government rejected the inclusion of CHP in the NFFO, electricity privatisation brought beneficial effects for CHP projects.[8] The newly privatised RECs, which were allowed to generate up to 15 per cent of their electricity supply, had an incentive to pursue economic schemes for local generation and management of peak demand. Two large RECs found opportunities to market CHP to industrial energy users (Harvey, 1994; Brown, 1994). CHP was also promoted under the EST's grant scheme. The CHP measure was thought to be one of the few measures which were well on course to meet their contribution to the targets under the *UK Programme*. By the end of 2000, however, the installed capacity was about 4,700 MW, approaching but not meeting the target of 5,000 MW.

A carbon/energy tax
The politics of a carbon tax – at both the EU level and the national level – revealed a clear gap between rhetoric and reality in global warming policy under the Conservative governments. Since the 1990 environment white paper, the government had repeatedly

stated its commitment to the use of economic instruments or market-based instruments in environmental policy. The establishment of this new principle was largely down to the Environment Secretary, Chris Patten, and his part-time adviser, David Pearce, an environmental economist. Pearce convinced the other government departments of the advantages of economic instruments and Patten championed the idea (interview with David Pearce, 1998). The use of economic instruments was also posited as a principle in developing the national climate strategy.

However, a carbon/energy tax was repeatedly rejected by the government. Patten's attempt to introduce a carbon tax in the 1990 white paper was thwarted by several economic ministries, in particular the Treasury, DTI and DEn. They were concerned about the effects of such a tax on their own policy objectives, such as keeping down inflation, the promotion of economic competitiveness and the electricity privatisation programme.

Britain was likewise firmly opposed to an EU-wide carbon/energy tax, which had been first proposed as a plank of the EU climate strategy in October 1991. On the grounds that the subsidiarity principle should be applied, that revenue recycling for energy efficiency measures as proposed might be inefficient and that the clause which made the EU tax conditional on similar measures being taken in other OECD countries was ambiguous, Britain resisted the EU proposal, together with France and four less developed member states. Given the unanimity voting procedure on taxation issues, their opposition led to the formal abandonment of the tax proposal in December 1994.

The EU tax could have been the central tool with which to achieve the British carbon dioxide stabilisation target. According to *Our National Programme*, it was estimated that the tax alone could reduce emissions by 7–8 million tonnes of carbon with relatively little economic impact (DOE, 1992: para. 8.14). Why, then, was the EU carbon/energy tax rejected? The main reason was that no major actors in the government supported the proposed tax. The DTI argued that there would be a negative effect on industrial competitiveness and the economy, in particular given the larger primary energy sector and the higher proportion of non-European trading that Britain had compared with other member states. The Treasury did not want an extension of EU interference into national taxation policy, and the DOE did not see the need for the tax in the form and at the level proposed by the European Commission. Moreover, there was an authoritative conviction that the EU tax was not the solution for Britain. The report on the EU carbon/energy tax in 1992 made by the House of Lords Select Committee on the European Communities opposed the tax owing to concern about its effectiveness, the difficulty of implementing it, employment loss in the coal and heavy industries, its macroeconomic consequences, and the impact on British competitiveness. The problem of competitiveness, especially, constituted 'a major objection' to the tax (House of Lords Select Committee on the European Communities, 1992: para. 107).

Government concern about the effect on British industry was also very strong. The following statement by Chancellor Clarke during the parliamentary debate indicates this:

> A carbon energy tax – an alternative [to VAT on domestic energy] favoured by the Liberal party usually and the Labour party often – would impose not only a tax on domestic fuel but extremely heavy costs on British industry, and would disadvantage us in world markets…. We shall rescue this country from such a tax, whether the idea comes from Europe, the British Liberal party or the British Labour party. (*Hansard*, 1994: col. 1053)

Whether such a tax would really undermine the competitiveness of British industry was a controversial issue. In its evidence to the enquiry on the EU carbon/energy tax by the House of Lords Select Committee on the European Communities (1992: para. 40), the oil company Shell, which carried out modelling work based on 'pessimistic assumptions', according to *ENDS Report* (207, 1992), admitted that the impact of the proposed tax would be insignificant except for in the metals sector.[9] FOE also emphasised the efficiency gains from the tax and argued for the opportunity to improve competitiveness that would arise with the introduction of a carbon/energy tax. However, there was strong opposition from the CBI, the Trade Unions Congress and industry, especially from the coal and energy-intensive sectors, and the competitiveness concern, in particular, prevailed in the policy debate.

And still on course towards the target

As we have seen, many of measures in *The UK Programme* faced problems at the implementation stage and this compelled the government to revise its carbon dioxide savings projections (see

Table 7.5 *Original and revised projections[a] for savings of carbon dioxide (in millions of tonnes of carbon)*

Policy/measures	Original target	Revised projections			
		2000	2005	2010	2020
VAT on domestic fuels	1.5	0.4	1	1	1
EST	2.5	0.5	1	1	1
Energy Efficiency Best Practice Programme	5	3.5	4	4	4
Road fuel duties	2.5	3.6	5	6	7
Public sector reductions in emissions	1.0	0.8	1	2	2
Building regulations	0.25	0.25	n/a	n/a	n/a
Minimum standards for electrical appliances	0.5	0.35	n/a	n/a	n/a
Increased renewable target	0.5	2	2	3	3
Increased CHP target	1	3.5	4	5	7

[a] Original target as set out in *The UK Programme* (Her Majesty's Government, 1994a) and revised targets as set out in the *Second Report* (Her Majesty's Government, 1997b: 53). See also DOE (1995: 25).
Note that these figures allow overlaps.
n/a: not applicable.

Table 7.5). Nevertheless, Britain was expected to meet the 2000 target with ease. This was largely a result of what was called the 'dash for gas'. Freed from the shackles of state interference, privatised electricity generators went for cheap gas instead of coal. Gas-fired power plants were also an economic means of meeting the requirement to limit emissions of sulphur dioxide, nitrogen oxides and dust under the EU Large Combustion Plant Directive, which otherwise would necessitate the expensive retro-fitting of coal-fired plants with flue gas desulphurisation (Collier, 1997). The shift to new gas-fired plants from other energy sources was 30–50 per cent higher than initially expected (DOE, 1995: 3).

Conclusions

The policies and measures that were adopted to reduce carbon dioxide emissions in Britain contrast with those in Japan. British policy-makers placed great importance on a non-interventionist approach and voluntary action by industry. Voluntary action was also emphasised by Japanese policy-makers, but in contrast to the Japanese policy framework, which was based largely on the combination of financial incentives/assistance and administrative guidance, the central feature of the British policy framework for voluntary action was the provision of authoritative information. What subsidies there were were available only for organisations with limited resources. Also, voluntary action in Britain was basically at the individual level; collective action orchestrated by trade associations, as seen in Japan, too often failed. The British hands-off approach was also seen in the new energy promotion policy, the NFFO, which was essentially a 'market-enablement' approach. All these British policy features fit with the institutionalists' expectations of policy contents.

However, to put too much emphasis on institutional influences is misleading, because this would obscure the strong influence of the Conservatives' pet policy of electricity privatisation on global warming policy. To what extent the hands-off approach is due to the British non-interventionist tradition or the Conservatives' party ideology will be discussed in the Epilogue. Here I simply point out that the effect of pluralist, non-interventionist institutions on global warming policy in Britain was reinforced by party ideology.

Another contrast with Japan's policy processes is the quick decision-making in setting targets. In contrast to the prolonged conflict between the EA and MITI over a specific target, in Britain, once Prime Minister Thatcher had decided to set a target, the government did so quickly. Similarly, when the original target for 2005 was brought forward to 2000, it was done so quickly under Prime Minister Major's initiative. In Britain, power is centralised and leadership is strong. These institutional characteristics enabled Britain to change policy quickly.

A related issue here concerns the role of individuals. In Britain individuals often catalysed policy developments. Apart from Thatcher and Major, Environment Secretary Patten and his adviser, Pearce, especially, had important roles in providing the idea of using economic instruments, and this idea could be entered into the central locus of policy debate. Also important was the role of influential politicians in 'reverberating' international influences within domestic politics (Putnam, 1988). They did so by marrying international messages or pressures with their own institutional power. The perceived international pressures on the British government to bring its original 2005 target forward were brought into domestic politics by Major. The message of the epistemic community on global warming similarly reverberated in British domestic politics through Thatcher. To put it other way round, led by a British scientist, the epistemic community had the political leader's ear. A contrasting case was the EU carbon/energy tax. With no institutionally powerful political actor advocating the idea, its lack of reverberation in British domestic politics was inevitable. Madison and Pearce (1994: 24–5) conjecture that the striking consensus against the EU carbon/energy tax within the government stemmed partly from its linkage with the issue of the expansion of the EU's jurisdiction. The government having experienced great difficulty over its ratification of the Maastricht Treaty, the EU proposal was too sensitive a political issue. If so, what happened was 'negative reverberation'. Issue linkages at the international level hampered domestic policy change. As seen in British climate strategy, the EU's role in climate policy was rather limited. Nevertheless, the EU did have significant influence on the British climate policy debate and actual policy, especially through its proposal for a carbon/energy tax.

While the British government respected climate science, this did not necessarily mean that Britain acted on the claim of the epistemic community. Environmental rationality was often overwhelmed by the perceived economic rationality. The EST's financial crisis is a case in point. Preoccupied with the creation of free markets, the government shied away from decisively intervening in the markets for environmental interests.

The British approach to global warming was weakened by economic interests. The government's policies of introducing both full-rate VAT on domestic energy (which it in fact failed to implement) and the fuel escalator instead of a carbon tax (in which it did succeed) were a result of intensive lobbying on the part of energy-intensive industries (which would have been hit hardest by the carbon tax), but while those industries successfully defended their interests the costs were passed on to the wider public. In the case of VAT, moreover, the public, in turn, partly defended its interests from the policy, which essentially benefits future generations. Here the issue-based approach gives some analytical insight. The imbalance of political power between the energy-intensive industries, the general public and future generations which arises from their different capabilities for collective action was combined with pluralist political processes (based on competition) with the result that the costs fell on those least capable of collective action. Moreover, the failure to introduce the full rate of VAT almost bounded the future policy path: as its effect on the EST would later indicate, it created difficulties for future policy developments involving price increases in domestic energy, however well intended.

Notes

1 'Equitable burden-sharing' was implicit in the conclusion of a joint Council of Energy and Environment Ministers in October 1990. 'Burden-sharing' would allow less developed EU countries to increase their emissions of carbon dioxide as their economies grew. The British government opposed this idea on two grounds. First, it implied a top-down approach; that is, the European Commission would be given discretion to decide policies from above. Second, there were concerns over the additional costs arising from the further reductions in emissions required to compensate for the increases in the emissions from these countries. In March 1993, at the Environment Council, the then junior Environment Minister, David Maclean, said that Britain was 'totally opposed to taking anyone else's burden' and that he had no intention of asking the British people to take on others' targets. See *ENDS Report* (218, 1993: 35).

2 As a result of a ruling by the European Court of Justice of 21 June 1988, VAT was imposed on non-domestic fuel and power from 1 July 1990.

3 The common definition of a fuel-poor household is one that needs to spend in excess of 10 per cent of household income to maintain a satisfactory heating regime.

4 The original plan for electricity was delayed, and the full introduction of competition in domestic markets had to wait until May 1999.

5 Although low profile, by 1998 the Energy Efficiency Standards of Performance scheme was widely judged to have been a success, including by the National Audit Office. See House of Commons Environment Audit Committee (1999a: para. 77); House of Commons Environment, Transport and Regional Affairs Committee (2000a: para. 46).

6 After the EEO was transferred to the DOE, it was reorganised as the Environment and Energy Management Directorate. However, throughout the present book I will use EEO to avoid confusion.

7   Combined heat and power includes co-generation, urban heating and cooling systems.

8   Waste-fired CHP was later made eligible for NFFO; however, because the majority of CHP was fuelled by gas or oil, most schemes did not receive the special treatment available under the NFFO.

9   The modelling work by Shell UK, however, suggested that a competitiveness problem might arise in the metal sector. Nevertheless, *ENDS Report* (207, 1992: 14–16) points out the pessimistic assumptions built into Shell's model. See also House of Lords Select Committee on the European Communities (1992: para. 43).

8

Competition and pressure: British policy integration

Britain has sought to integrate environmental concerns into policy decision-making at all levels. To this end, the first environment white paper introduced two institutions which would 'ensure that ... environmental issues are fully weighed in decisions'. One was the Cabinet Committee for the Environment, chaired by the Prime Minister. This was later replaced by the Ministerial Committee on the Environment, chaired by the Leader of the House of Lords instead of the Prime Minister. The other was the introduction of a 'green minister' in each government department. In addition, the white paper set up two institutional mechanisms to facilitate policy integration: an environmental section in departmental annual reports and environmental appraisal of policies at an early stage in their development by departments (Her Majesty's Government, 1990: paras 1.6–1.7). In the first white paper on sustainable development, published in 1994, the government confirmed its commitment to policy integration (Her Majesty's Government, 1994b: para. 29.2). How and to what extent was this government aspiration achieved over the problem of global warming?

Energy policy and global warming

In contrast to Japan, Britain is an energy-rich country, with indigenous reserves of coal, oil and natural gas. In the postwar era, and especially after the oil crises, Britain pursued the development of a range of indigenous energy sources to ensure a sufficient energy supply to meet the maximum conceivable demand (O'Riordan *et al.*, 1988; McGowan, 1993). There was, however, no long-term, strategic energy policy. Before privatisation was undertaken in the energy sector, the British energy policy was characterised by a 'haphazard process of piling measure on measure' – measures which were instant responses to pressing problems made with regard to short-term political considerations (McGowan, 1993; Robinson, 1993). The main tool for pursuing such a policy was the publicly owned energy industries, especially the Central Electricity Generating Board (CEGB), a statutory monopoly. The DEn, the CEGB and regional electricity boards constituted a corporatist, vertically integrated policy network (Roberts *et al.*, 1991: 39–43). Using this network the government had protected coal and nuclear energy. Coal, in particular, accounted for about 80 per cent of fuel requirements before privatisation.

In the 1980s, however, the Conservatives' philosophy of a free market with minimum state interference hit the energy sector. Thereafter, establishing competitive, free energy markets and minimising market intervention were the central thrusts of energy policy in Britain. Over a decade from the mid-1980s, the oil, gas, electricity, nuclear energy and coal industries were privatised, with the biggest project among them being the privatisation of the electricity industry. As already indicated in Chapter 7, the integration of the concern with global warming into energy policy was profoundly affected by the government's preoccupation with successful electricity privatisation and the introduction of free, competitive markets. This created a severe test of the government's ability to reconcile neo-liberal ideology with essentially interventionist environmental policy.

Competitive prices versus energy efficiency

In addition to the traditional economic and social arguments (the latter relating to the problem of fuel poverty), the problem of global warming gave a new case for energy efficiency at the end of the 1980s. A new energy efficiency campaign was promised by the then Energy Secretary, Cecil Parkinson, as soon as global warming became an important item on the political agenda, and the 1990 environment white paper stressed the need to promote energy efficiency as the means to curb emissions of carbon dioxide (Her Majesty's Government, 1990: 68–9). These initial moves were, however, largely a political gesture. The Thatcherite project sought to establish free competitive markets, and the successful sale of the electricity and other utility companies and liberalisation of the markets were priorities for the government.

The Electricity Bill, published just after Thatcher's well known speech that addressed the problem of global warming in 1988, had little to say about the environmental effects of liberalising the market – specifically in terms of energy efficiency. Not only did the proposed bill fail to contain an effective instrument to promote energy efficiency, but also the proposed market design would give the privatised companies an incentive to sell *more* electricity. Intense lobbying was begun by environmental groups, led by the FOE and the ACE, to have the bill place environmental duties on the regulator, the Energy Secretary and the privatised companies. The government was also urged by the House of Commons Energy Select Committee to take energy efficiency more seriously and to introduce in the privatisation proposal a regulatory mechanism to encourage electricity companies to invest in energy efficiency.

Energy efficiency became an important issue during the passage of the bill. The Lords passed an amendment, with all-party support, that imposed responsibility on the Energy Secretary to ensure that electricity suppliers promoted the efficient use of electricity. In the Commons, twenty-five Conservatives signed a motion calling for new powers for the regulator to set energy efficiency standards for electricity suppliers. The DEn strongly opposed to the move. However, having been pressured on energy efficiency by broad advocacy the government eventually accepted the need to incorporate a provision for energy efficiency. Given the technical difficulties in adopting the Lords' amendment, the government replaced it with its own alternative, which was weaker than the Lords' (Roberts *et al.*, 1991: 120–39).

The new provision enabled the regulator to determine, from time to time, standards of performance for energy efficiency that privatised electricity suppliers were required to achieve. Thus, detailed policy-making and implementation concerning the promotion of energy efficiency were left to the regulator. The regulator's primary duty was, however, to promote competition, and the promotion of energy efficiency was only a secondary duty. The regulator's other environmental duties were also very weak, with the Electricity Act requiring it only to 'take into account' the environmental impacts of the electricity business. Although the provision of standards of performance later proved one of the most successful measures under the British climate/energy policy, these duties of the regulator inevitably became the source of controversy.

While the successful privatisation of utility companies was the major reason for the government's reluctance to accept substantial legislative provision for energy efficiency, it also reflected the traditional British view. A reply of an official at the DEn to the House of Commons Energy Select Committee expressed this well:

> Energy efficiency measures are by definition in the economic interests of energy consumers.... The Government therefore starts from the position that consumers do not need subsidy, regulation or financial incentive or penalty (through the taxation system) to get them to act to improve their energy efficiency, as long as the market is working properly. (House of Commons Energy Select Committee, 1991: para. 40)

The implication is that there is a real and manifest built-in incentive in energy efficiency measures; government intervention is therefore unnecessary and the most effective policy is 'to encourage voluntary compliance' by giving information so that individual actors can choose the most cost-effective measures (House of Commons Energy Select Committee, 1989: para. 111). During the parliamentary debate on the Electricity Bill, the emphasis on the inherent economic advantages of energy efficiency, combined with a belief in market efficiency, led the government to argue that the competitive pressure introduced by privatisation would be the best means to encourage investment in energy efficiency.

After the general election in 1992, the DEn was abolished, and administrative responsibility for energy efficiency was transferred to the DOE and the rest to the DTI. This institutional integration symbolised a shift in the main aim of energy efficiency, from 'to increase economic efficiency in the economy as a whole' (DEn, 1983, quoted in Roberts *et al.*, 1991: 124) to environmental protection. Indeed, the DOE soon signalled the change in the government approach to energy efficiency. In the environmental white paper published in October 1992, the government admitted that:

> In the past, the Government has concentrated on providing information about cost-effective energy efficiency measures but it is unlikely that information programmes by themselves will realise the full scope for savings. (Her Majesty's Government, 1992: 53)

It reiterated the commitment since 1990 to the use of economic instruments, as follows:

> experience has shown that, without an external stimulus, voluntary action to improve energy efficiency tends to be limited ... [price measures] can be very effective relative to the alternative options – regulation or the provision of subsidies for investment in energy efficiency. (Her Majesty's Government, 1992: 45)

In 1993 the Environment Secretary, John Gummer, publicly stated that 'our traditional energy policies are not sustainable' and called for 'the right energy policy' (DOE, news release, September 1993). Also, Andrew Warren, Director of the ACE, was appointed as a special adviser to the House of Commons Environment Select Committee, thus providing energy efficiency advocates with improved access to the locus of policy debate.

The main policy instrument to implement this new approach was the EST, the establishment of which was an electoral pledge of the Conservatives. The government expected the EST to be the primary driver in promoting energy efficiency; this is clear from the fact that the EST was asked to produce 25 per cent or more of the total carbon dioxide saving required under the national climate strategy. The EST was an innovation which would solve the dilemma arising for the newly privatised gas and electricity companies: on the one hand they had been asked to promote energy efficiency, but on the other the new regulatory regime gave them an incentive to do otherwise (Manners, 1995: 150). The EST would also meet the contradictory political desires to respond to pressures for more interventionist policy and to minimise market intervention. As seen in Chapter 7, however, the initial expectation of the EST was shattered when it was proven that the politics of policy integration had been personalised as a result of a lack of clarity in the statutory duties placed on the regulators in terms of promoting energy efficiency, the wide discretionary powers given to the regulators, and the government's reluctance to interfere with the markets (Helm, 2003: 274–5).

The case of the EST, however, has further implications. The EST, in a sense, fell into the trap of the new politics of energy efficiency. The problem was institutionalised as a result of the government trying to integrate global warming concerns into energy policy. Before it was abolished, the DEn had been given the task of incorporating environmental concerns into energy policy, hence striking a balance between environmental and economic or competition objectives. However, the subsequent departmental reorganisation, by splitting the environmental aspect of energy efficiency (which went to the DOE) away from the economic or competition aspects (which went to the DTI), brought the discord between the two objectives to the surface. This evolved into a conflict between two adversarial coalitions. On the one hand, a new policy network emerged, composed mainly of the DTI, gas and electricity regulators and major energy generating and supply companies. They were predominantly concerned with energy supply, and together with energy-intensive industries were preoccupied with competitive energy prices. Their view was that market forces were sufficient to promote energy efficiency. As such, the appropriate role of the government and regulators was merely to ensure fair competition in the energy markets. For them, greater efficiency in electricity generation and the rapid growth in CHP installation after privatisation showed that enhanced competition promoted energy efficiency.

On the other hand, a loose policy network began to emerge involving the DOE, EST, environmental groups, the ACE and those groups engaged with the fuel-poverty problem. As seen below, this loose network was most prominent where social policy objectives were brought to the fore. In general, their view on energy efficiency was that more proactive government leadership and intervention were necessary for further improvements, because market barriers were higher than the economic theory of energy efficiency would suggest. The DOE was not necessarily at the forefront in this argument, but was more receptive to the idea that government should play a more proactive role in promoting energy efficiency. As explained in Chapter 7, Environment Secretary John Gummer became instrumental in maintaining proactive measures by making funding available for the EST. In particular, he put an emphasis on developing energy service companies, which the Director-General of Ofgas and the Energy Minister Tim Eggar claimed would be delivered through enhanced competition (DOE, news release, 11 May 1995; *ENDS Report* 244, 1995: 13). In the 1996 environment white paper, the DOE included among its priorities the development of the sales of energy services rather than the supply of energy alone (Her Majesty's Government, 1996: 9).

The EST lay uneasily between these two adversarial networks. While it had its genesis in the argument of the energy efficiency advocacy network, because of its institutional design the fate of the EST inevitably largely rested with the market-force network. The shift of administrative responsibility for energy efficiency to the DOE made the policy process more open, stimulating developments in policy debate on energy efficiency and giving opportunities for hitherto peripheral energy efficiency advocates to deliver their messages more effectively. At the same time, however, the separation of the two conflictual administrative responsibilities between the DOE and DTI posed institutional constraints on policy integration by making the process inherently conflict-ridden. Institutional integration was double-edged.

In the end, the government prioritised the ideology-based political strategy of drawing back the state from markets, while, at the same time, dismissing the existence of the conflict between the two. In the *Energy Report 1995*, subtitled *Competition, Competitiveness, and Sustainability* (DTI, 1995b), the government endorsed the argument that competitive markets and environmental objectives are two sides of the same coin, since they are both essentially about making the best use of resources. An obvious shortcoming of this argument is that it evades the problem of how to motivate consumers to switch to a more energy efficient form of consumption when the liberalised energy markets were increasingly price competitive. The government was pressured by environmental groups and various advisory bodies into addressing this problem; however, it basically continued to depend on the built-in economic opportunities of energy efficiency. The government's attitude was criticised by the UK Round Table on Sustainable Development (1997: paras 3.21, 4.4) as 'extremely' disappointing.

One area where energy efficiency advocacy was making progress was the problem of fuel poverty. Although the government was strongly committed to the creation of the free energy markets, the sheer scale of this social problem warranted some market intervention to promote energy efficiency. Thus, two of the few energy efficiency measures beyond information provision, the Energy Efficiency Standards of Performance scheme and the HEES, were basically part of social policy (Helm, 2003: 348).

The importance of the social dimension of energy efficiency was rediscovered with the failure to introduce full-rate VAT on domestic energy, which ironically triggered policy developments on energy efficiency in the form of a number of private members' bills submitted to Parliament. Among the notable successes was the passing of the Home Energy Conservation Act 1995, which required local authorities to take action on energy efficiency. Through two (initially unsuccessful) attempts to pass the Home Energy Conservation Bill, a broad coalition was built up involving environmental groups, those concerned about poverty, pensioners and health problems, the energy efficiency industry and the construction industry.

The Treasury's concern over expenditure implications was often a major stumbling block for the various private members' bills introduced. A notable case was the Energy-Saving Materials (Rate of VAT) Bill, which would have corrected a tax distortion whereby full-rated VAT applied to energy-saving products and services while half-rated VAT applied to domestic energy. Like the Home Energy Conservation Bill, it created a broad coalition of support involving the CBI, the Builders Merchants Federation, chambers of commerce, pensioners' rights groups, FOE and the ACE. Although the Treasury had been opposed to the bill, by the time a third attempt was made in 1997 parliamentary support had built up and it was expected to be passed in both Houses. However, the Labour Party made an abrupt policy U-turn and this buried the bill. Labour was allegedly concerned about the implications of the measure for revenue, and the party anticipated coming into office after the forthcoming general election. The shadow Chancellor had already promised to cut VAT on domestic fuel from 8 per cent to 5 per cent, leading to a £450 million reduction in the annual budget.

Nevertheless, the number of such private members' bills suggests that the problem of fuel poverty, while certainly a constraint on the integration of the climate consideration into energy policy in terms of raising energy prices, was also an important driving force for policy integration in Britain.

Renewable energy

It was a common view throughout the spectrum of British politics that Britain had failed to harness the world's largest resources for such renewable energy as wind, wave and tidal power. After the first oil crisis, the government launched large-scale energy research, development and demonstration (RD&D) programmes by establishing the Energy Technology Support Unit (ETSU). The result was a 'very expensive disaster' (interview with Michael Massey, 1998). After Thatcher came into office, expenditures for R&D into renewable energy were drastically cut. This history of British policy on R&D on renewable energy was described by the House of Commons Energy Select Committee as follows:

> In the past, the Department [of Energy] has attempted to establish costs at too early a stage in a technology's development; to draw final conclusions from tentative assessments; and placed too little value on continuity of funding. It is difficult to regard the history of renewable energy R&D funding in the UK as other than a history of volte-face, premature judgements and plain errors. (House of Commons Energy Select Committee, 1992: para. 55)

Electricity privatisation at the end of the 1980s gave a boost to those renewable energy sources that were close to the point of commercialisation. Facing the dilemma of the Thatcherite commitments to freeing energy markets and to protecting and promoting nuclear energy, the government devised a nuclear obligation and a nuclear tax whereby the privatised RECs would be required to buy a specified amount of nuclear capacity, thus insulating nuclear energy from market forces. After intense pressure from the advocates of renewable energy, however, the government refined the scheme to include the NFFO and the fossil fuel levy, by preparing a separate protected market for a total of 600 MW over ten years for renewable sources. Although the government justified the scheme in terms of security and diversity of supply, with only a supplementary reference to global warming, the inclusion of renewable sources in the scheme was 'more a matter of "spin" than deliberate policy choice' (Helm, 2003: 350). The extent to which the renewable component was an afterthought can be seen from the use of the revenues raised from the fossil fuel levy: none at all went to renewable sources of energy in 1990, and even until privatisation of the nuclear power industry it had risen only to about 10 per cent (Mitchell, 1998). Nevertheless, with the NFFO the government had for the first time a policy to promote the use of renewable energy. After the second order of the NFFO in 1991, partly to ease the administrative process of the NFFO exercise and partly to prevent only a few cheaper technologies benefiting from the scheme (Roberts *et al.*, 1991: 152), the DEn introduced the 'technology bands' or sub-quotas for specific sources of renewable energy, which would change in each successive order.

Developments in the policy on renewable energy since then indicate the increasing influence of global warming concerns. In 1990 the environment white paper established a target for the installation of 1,000 MW of renewable energy capacity by 2000. This was a leap from the 600 MW which the then Energy Secretary, Cecil Parkinson, said would be difficult to achieve with the NFFO (House of Commons Energy Select Committee, 1989). Aided by the great interest in renewable energy in the market,[1] a junior minister at the DEn, Colin Moynihan, further announced that the target would be reviewed, and described the 1,000 MW target as 'meaningless' (House of Commons Energy Select Committee, 1992: para. 13). Following major reviews of renewable policy by the Renewable Energy Advisory Group, in 1993 the government announced a new target of 1,500 MW by 2000.

The cost of generating electricity from renewable sources almost halved on average under the NFFOs. This success, however, must be balanced with the failure to achieve the target. Although 933 projects (representing 3,638 MW capacity) had been contracted for by 2000, only 371 were operational in June 2001, giving rise to only about 942 MW capacity. The stumbling block was planning permission. The larger wind farms were the worst affected by this: only 41 per cent of requests for planning permission under NFFO-3 to NFFO-5 were successful (Performance and Innovation Unit, 2001). The problem at hand was how to meet the national policy objective for renewable energy and at the same time to meet local environmental objections. Wind power is certainly a sustainable form of energy production, but it also causes environmental damage in the form of noise and damage to scenic beauty. The impact on landscape was especially a matter of concern, since the most profitable sites tended to be environmentally sensitive locations. Wind farms, thus, often provoked nature conservationists as well as local residents. Environmental groups were divided. Those concerned about nature conservation were cautious about advocating renewable sources of energy, and emphasised the importance of energy efficiency before resort was made to renewable energy. Under the British planning system, it is the job of local planning authorities to strike the balance between local and national environmental objectives. In 1993, to reduce uncertainty at the stage of planning permissions, the DOE published a planning guidance note on renewable energy (Planning Policy Guidance 22), which emphasised the contribution renewable sources of energy can make to carbon dioxide reduction, so as to urge local authorities to have more regard to national policy objectives. However, local, hence concentrated environmental interests not surprisingly had the upper hand over the much more dispersed global environmental interests.

Lack of a strategic view
While the developments of the NFFO scheme indicate the growing influence of the global warming consideration on energy policy, the long-term R&D policy for renewable energy indicates otherwise. The downward trend since the early 1980s in R&D spending continued in the 1990s and between 1987 and 1998 expenditure on R&D in renewable energy sources fell by 81 per cent. Under the Conservatives, the emphasis was very much on a short-term measure with the aim of giving a 'kick-start' to those technologies near commercial exploitation (that is, the NFFO), at the expense of a long-term RD&D policy, positing the latter as 'complementary' to the former. The government's R&D strategy was to rely on EU funds, despite the past failure to attract them. Public expenditure on RD&D for renewable sources were set to be cut over ten years, with a cut by 20 per cent to £14.7 million in 1996 as a start. Potential but still nascent technologies such as offshore wind power, tidal power and wave power were hit hardest. The Treasury was looking for opportunities to reduce government R&D expenditures drastically, and for the Treasury, which was traditionally suspicious of investments in projects which could not show tangible returns within the short to medium term (Weale, 1997: 106), these technologies were sound options. The cut was also in response to a critical review of the government's RD&D expenditures on renewable energy by the House of Commons Public Accounts Committee. British policy-makers were strongly concerned about 'wasted millions' in the failure to pick technological winners in the past (interview with Michael Massey, 1998).

Coal protection lingers on

An acid test for policy integration centres on the influence of environmental concerns on the long-standing privilege given to coal. Electricity privatisation was a good opportunity to remove the coal protection regime. Indeed, it was a 'hidden agenda' of electricity privatisation (Roberts *et al.*, 1991: 57–77). However, the outright removal of protection would have made privatisation itself difficult; the government then came up to the solution of three-year special contracts between the coal industry and the privatised electricity generators, on the ground that this would provide the coal industry with time to adjust to market forces, thus leaving the difficult political issue to be tackled after the next general election.

In October 1992, just before the three-year post-privatisation protection expired, the re-elected Conservative government announced the closure of thirty-one pits out of fifty within six months. This provoked a strong public reaction. The President of the Board of Trade (Trade and Industry Secretary), Michael Heseltine, quickly announced a moratorium on most of the announced closures and started a policy review. The review was concluded in March 1993 and in a white paper the DTI announced a new pits 'rescue' plan over the period until 1998, the deadline for the next general election. A new contract was to be in place between the coal industry and the power generators for higher volumes of coal at higher prices than otherwise and the government was to provide a further subsidy for generators to burn extra coal. This was a generous rescue plan for the coal industry, given the increasing pressure on it emanating from cheaper gas and imported coal, and the improved performance of nuclear power (Helm, 2003: 177–9). The white paper, however, made clear the government's intention to privatise the industry, and hence retreat from interference with the markets for coal. The coal industry's demand to constrain the number of new gas-fired power generation plants was also rejected in defence of free, competitive energy markets. Moreover, it was also stated that, in the long term, the problems of sulphur dioxide and carbon dioxide emissions would have to constrain the amount of coal that could be burned (DTI, 1993: para. 2.42). Thus, in the long term government advanced its preference to be freed of the coal shackle; however, in the short term the industry effectively defended its interests.

According to Skea (1993), even a partial rescue of the coal industry would have largely cancelled the expected carbon dioxide savings from other measures, such as the EST or the road fuel duty escalator. In the 1990 environment white paper, the government noted that 'Changes in the pattern of energy use can cause problems and expense for some individuals, some industries and some countries, but are necessary if global warming is to be avoided', and argued for the need for a public debate in making difficult decisions (Her Majesty's Government, 1990: paras 1.34–35). There was little public debate on the effect of the coal rescue policy on the environment, however. Moreover, given the sensitivity of the issue, environmental groups were also, in Heseltine's words, 'extraordinarily quiet' (see House of Commons Environment Select Committee, 1993). The House of Commons Environment Select Committee (1993) was acutely critical of the lack of debate, arguing that the environment was at best an 'afterthought' in making the new policy.

Transport policy and global warming

British transport policy was, for decades, formulated on two key assumptions: first, private car use would increase as income grew, and therefore it was necessary to expand road capacity; and second, public transport would decline, and therefore government expenditure on it would be reduced (Rawcliffe, 1995: 29–30; Goodwin, 1999). Rapid motorisation at the expense of other transport modes by proportion, as shown in Figures 8.1 and 8.2, is not a surprising outcome. During the 1980s the tendency towards road transport and away from public transport was accelerated under Thatcher, who declared she would do nothing to stop the 'great car economy'. The car symbolised positive things like 'progress', 'freedom of choice' and 'self-improvement' (Whitelegg, 1989). When the problem of global warming came onto the national agenda, the transport sector accounted for about 24 per cent of total carbon dioxide emissions in Britain (Figure 8.3), and more than 90 per cent of these emissions were from road transport. Also, carbon dioxide emissions from road transport were expected to contribute to most of the predicted rise in the total carbon dioxide emissions in Britain (Table 8.1). The problem of global warming posed a great challenge, entailing an about-turn in established, car-biased transport policy.

Figure 8.1 *Trends in passenger transport by mode, 1952–2001. 'Other' road includes motor cycles and pedal cycles.*

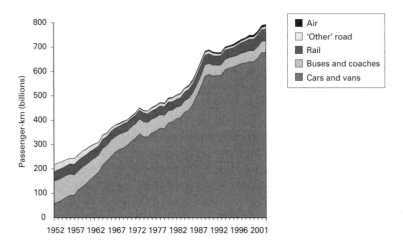

Source: DOT (2004).

Figure 8.2 *Trends in freight transport by mode, 1953–2001.*

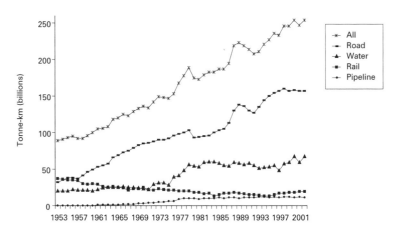

Source: DOT (2004).

Figure 8.3 *Carbon dioxide emissions from the transport sector, 1992 and 2002.*

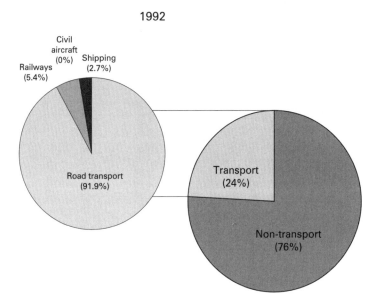

1992

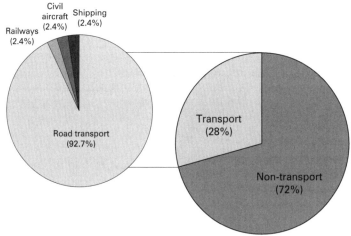

2002

Source: DOT (2004).

Table 8.1 *Forecasts of British carbon dioxide emissions (millions of tonnes of carbon)*

	1995	2010	2020	*Percentage change, 1995–2020*
Forecasts based on low fuel prices				
Transport	35	42	47	+34.3
Rest of the economy	115	106	110	– 4.3
Total	150	148	157	+4.6
Forecasts based on high fuel prices				
Transport	35	41	46	+31.4
Rest of the economy	115	108	109	−5.2
Total	150	149	155	+3.3

Source: DOT (2004).

Improving fuel efficiency

When the government started to discuss the British policy response to global warming, transport, and in particular road transport, was singled out by some departments as the main sector to bear the burden of reductions in emissions of carbon dioxide, not least because it was the source of the fastest growth in emissions (Madison and Pearce, 1995). The DOT, however, firmly declined to be the focus for the carbon dioxide reduction policy and refused to set targets for either energy consumption or emissions in its sector, or to establish regulatory measures to improve fuel efficiency, which it thought would damage the competitiveness of the car manufacturing industry. The 1990 environment white paper indicated the government's reluctance to interfere with the 'great car economy' as a means of tackling global warming, and emphasised instead voluntary action (Her Majesty's Government, 1990: 72–4).

The lack of clear objectives and policy within the DOT on transport energy use was repeatedly criticised by the House of Commons Energy Select Committee (1989: para. 121; 1991: para. 132). The Committee (1991: para. 131) also suspected that the DOT used the policy objective of improving energy efficiency to justify its massive new road-building programme, *Roads for Prosperity*, announced in 1989, which the Committee believed would result in more fuel consumption due to new traffic. On fuel efficiency policy the DOT basically passed the buck to the Treasury, on the grounds that improvements in fuel efficiency would best be pursued through the use of fuel duties (House of Lords Select Committee on Science and Technology, 1996). In fact, the inter-departmental group of economists which took the lead in the early stages of formulating global warming policy reached a unanimous recognition of the relative advantage of market-based instruments and especially the role of energy prices in reducing carbon dioxide emissions (Madison and Pearce, 1995: 126–7). Britain had also had success in using fuel duties to tackle the problem of air pollution arising from the use of leaded petrol.

An early major initiative came at the European level. In 1991 the European Automobile Manufacturers' Association announced that it would work towards a sales-weighted reduction of 10 per cent in carbon dioxide emissions from new cars in the EU between 1993 and 2005. The then Transport Secretary, Malcolm Rifkind, criticised this plan for being too unambitious and warned the British motor industry to aim for a 40–50 per cent improvement in fuel efficiency by 2005, which was later supported by the Royal Commission on Environmental Pollution. The Advisory Committee on Business and the Environment also urged the Society of Motor Manufacturers and Traders to move the target forward. The motor industry at both the EU and the British levels was fiercely opposed to the call, arguing that it was unrealistic.

A breakthrough came in 1993, when the then Chancellor, Norman Lamont, promised to introduce the fuel duty escalator in his March budget, as explained in Chapter 7. Although there was strong opposition from motoring organisations and the haulage industry, Lamont's successor, Kenneth Clarke, was not only committed to Lamont's initiative, but strengthened the measure by increasing the rate of annual increase in fuel duties. As noted in Chapter 7, the measure was politically expedient. The Treasury had a macroeconomic strategy of shifting the source of revenue from direct to indirect taxes and also wanted to reduce a huge public deficit. When Lamont announced the introduction of the fuel duty escalator, he predicted an overall budget deficit of £35 billion for 1993 and £50 billion for 1994, representing 8 per cent of GDP. After he took over as Chancellor, Clarke, in his first budget, pledged to eliminate the public deficit by the end of the

1990s and planned to raise £10.5 billion over three years from tax increases and spending cuts. The fuel duty escalator, which was to raise £2 billion by 1996/97, was an 'essential part' of his plans to deliver healthy public finances. The visibility and the real effect of this measure were lessened by falls in world oil prices. Combined with the low profile given to the measure in public debate and a freeze in vehicle excise duty for road hauliers, fuel duties represented a good source of revenue with low political cost at the time.

The government's commitment to the fuel duty escalator was welcomed by environmental groups and authoritative bodies, including the RCEP. Meanwhile, they also continued a call on the government to reform motoring taxes further, so as to strengthen incentives to improve fuel efficiency. One focus of these calls was vehicle excise duty, which was set at a flat rate regardless of fuel efficiency; another was the tax on company cars, as it encouraged car use by those who had them, because they were taxed less for driving longer distances and had their fuel subsidised through tax breaks. Although company cars accounted for only around 8–10 per cent of cars in Britain, they were responsible for about 20 per cent of car traffic (RCEP, 1994: para. 6.7). The RCEP also warned the government that a doubling of fuel prices by 2000, amounting to a 9 per cent annual increase, would be necessary to achieve the objective of carbon dioxide stabilisation by 2000, and called for further use of fuel duties to promote cleaner fuels, such as compressed natural gas (CNG) and liquefied petrol gas (LPG).[2]

Clarke cautiously, but gradually, started to incorporate environmental concerns into motoring taxes in his budgets. During the process of radical policy change outlined below, there was established a new principle of reflecting the full costs of transport into prices. The budget of November 1996 presented a large package of fiscal measures to tackle environmental problems caused especially by vehicles. In the budget statement – the last budget under that series of Conservative administrations – Clarke said that 'motorists should bear the full costs of driving – not only wear and tear and congestion on the roads, but the wider environmental costs' (*Hansard*, 26 November 1996: col. 167). This reflected one of the main conclusions of the green paper on transport published earlier in the year in response to the RCEP's critical report on transport policy, which pointed to 'misleading price signals', and to the 'great national debate' on transport policy launched by the then

Transport Secretary, Brian Mawhinney. The focus of Clarke's green measures was the problem of air pollution; this was a response to mounting public concern over the effect of traffic exhaust on health and new policy initiatives introduced at both the domestic and European levels (respectively the national air quality strategy being developed by the DOE and the EU framework directive on ambient air quality and assessment adopted in 1996).

While welcomed by environmental sympathisers, Clarke's green measures were far short of what his bold statement actually implied, especially when looked at from the imperative of improving fuel efficiency. Although the Chancellor did not give in to pressure from motoring organisations to abolish the fuel duty escalator, neither did he act on the RCEP's recommendation to double petrol prices by 2000. The long-standing call for reforms to company car taxes and vehicle excise duty was also ignored (RCEP, 1997: paras 6.7 and 6.8). Again, motoring organisations and, in particular, car manufacturers, fearing the impact on domestic demand for their more profitable larger cars, were opposed to such measures. Indeed, the Chancellor's green measures should be seen in the context of the forthcoming general election. The environment had been placed increasingly higher up on the Labour Party's agenda and the Liberal Democrats went even further, in advocating a wholesale green tax reform. While earning green credentials with his landmark package for the integration of environmental costs into motoring prices, the Chancellor, or more broadly the Conservative Party, was also careful not to provoke motorists any further.

R&D on LEVs: the British non-interventionist approach
A lack of interest in LEVs was a general political and industrial feature of Britain, condoned by the House of Commons Energy Select Committee (1989), which gave short shrift to the potential of LEVs in terms of reducing emissions of carbon dioxide. The government's interest in LEVs came only around the mid-1990s, with a push from the RCEP and in parallel with the growth in interest in LEVs within the EU. The main measures included: a price differential in favour of CNG and LPG; the Powershift Programme; and the Foresight Programme. Powershift was a three-year programme launched in 1996 and operated by the EST. It was designed to encourage fleet operators to switch to

'clean' fuels such as CNG, LPG and electricity. Nevertheless, only £6 million in total was allocated for the three-year programme. This compares poorly to the more than £12 million equivalent spent on LEV promotion by MITI in the fiscal year 1996 alone. The Foresight Programme was a British innovation to 'pick winners'. It contained sixteen sector panels, including one looking at R&D into renewable sources of energy. Recognising the need to take action to nurture the right, viable technologies but unwilling to get involved in the business of picking winners, the government created a locus for collaboration between academia, business and the science base within government to identify and promote 'winners'. It was an 'industry-led, market-driven' programme. As an official at the DOT said, the government was 'dependent on the industry in taking up new technology and offering it to the consumer'; the government's role was 'to facilitate and stimulate, not to lead the development of technology' (House of Lords Select Committee on Science and Technology, 1996). The RCEP (1997: paras 2.88, 8.13–18) argued that forcing the development of vehicle technology was an essential measure to reduce fuel use to the necessary level on environmental grounds and urged the government to intervene in the markets more proactively. The government, however, chose essentially to privatise policy-making in this sphere.

Traffic restraint

During the 1990s road policy underwent a dramatic 'paradigm' shift. The shift was not necessarily induced by global warming concerns alone, but had significant implications for carbon dioxide emissions in the transport sector. The root cause was the publication of two documents by the DOT in 1989. One was a revised national traffic forecast which predicted an increase in traffic of 82–134 per cent between 1988 and 2025 (DOT, 1989a). The other was the DOT's response to this forecast (DOT, 1989b). The DOT launched a massive road construction programme, *Roads for Prosperity*, which, when added to the existing programme, would more than double the total trunk roads programme. This response was a corollary of the traditional 'predict and provide' approach to road transport policy. The main assumptions behind the approach were that the total volume of traffic was closely related to income and largely unaffected by policy; therefore, road capacity should be expanded to match the increase in traffic volume (Goodwin, 1999).

The new road scheme was claimed to ameliorate the problem of congestion, the cost of which, according to the CBI, amounted to around £15 billion a year or about 2 per cent of GDP (CBI, 1989). It was also a 'solution' to the growing public concern over the health hazard of vehicle emissions, because new roads would reroute unsuitable traffic away from towns and better traffic flow would reduce emissions of pollutants. The massive road programme, which pleased the CBI, the British Road Association and the road haulage industry, however, was quickly criticised by the DOE. The Environment Secretary, Chris Patten, who at the time was preparing the first environment white paper, criticised the traffic forecast as 'unacceptable'. The DOT responded that the forecasts were not a target, and in 1991 the Transport Secretary, John MacGregor, further said that 'There will be cases … where on economic or environmental grounds, or indeed both, it is neither practicable nor desirable to meet the demands by building the roads' (Whitehead, undated). However, this, of course, did not mean that the DOT gave up implementing *Roads for Prosperity*. Indeed, it was a priority for the DOT.

The group most provoked by *Roads for Prosperity* was nature conservationists. According to English Nature, as many as 150 designated sites of special scientific interest, compared with the DOT's estimate of 48 sites, were potentially at risk from the road programme (RCEP, 1994: 58). Some cases were brought to the European Commission by anti-roads groups. The British government was publicly accused of failing to comply with the EU's legal requirement to undertake an Environmental Impact Assessment before any road-building project and threatened with legal action (Dudley and Richardson, 2000: 150–62). More crucial in damaging the implementation and credibility of *Roads for Prosperity* were the growing anti-road protests, based on direct action on the road construction sites by an unusual social alliance of counter-culturalists, established environmental groups and the local residents directly affected by new roads.[3] In particular, between 1993 and 1996, anti-road campaigns at various places in Britain attracted wide media coverage and mobilised public sympathy, and brought about some change in public views regarding roads. The emerging national sentiment inspired politicians to set up an anti-road parliamentary group in 1994. According to the FOE, as many as thirty-five Conservatives held marginal seats that would be affected by proposed trunk road schemes (*The Economist*, 28 January 1995; *Financial Times*, 19 October 1994).

In 1994, the DOT made a first major concession in *Sustainable Development: The UK Strategy*, published in January (Her Majesty's Government, 1994b). The DOT acknowledged the pressing need to influence the rate of traffic growth and road transport needs (paras 26.20, 26.44) and this new policy was embodied in the form of the planning policy guidance on transport, a product of collaboration between the DOT and DOE aimed at integrating road-building policy into land-use policy so as to reduce the need and demand for the use of cars. The DOT also started a 'package approach' to funding local transport, in which priority was given to those bidders who envisaged measures to promote alternative transport modes to car travel or otherwise to manage car use.

More important challenges to the dominant roads policy discourse, however, came in the form of two authoritative reports. One was by the RCEP, on the environment and transport (RCEP, 1994). The RCEP criticised existing transport policy as 'unsustainable' and made over 100 recommendations to shift the 'predict and provide' approach to a 'reduce car dependency' approach. Among the recommendations were a halving of the expenditure on trunk roads, greater support for public transport, greater integration of land-use and transport policies, establishing targets to cut carbon dioxide emissions from transport and a doubling of petrol prices. The other report was a study by the Standing Advisory Committee on Trunk Road Assessment (SACTRA) for the DOT. In the report, SACTRA admitted that induced traffic 'can and does occur, probably quite significantly', and that this reduced the duration of congestion relief by means of road expansion (SACTRA, 1994: para. 15.04). The traditional approach and values in road transport policy were, thus, authoritatively rejected.

The Treasury immediately and consecutively cut public spending on road-building. Between 1994/95 and 1997/98, public spending on trunk roads in England fell from more than £2 billion a year to less than £1.5 billion a year (RCEP, 1997: para. 6.10). This was a drastic policy shift. The more sincere response to the transport crisis was the launch of the 'great debate' on transport policy in late 1994 by the then Transport Secretary, Mawhinney. Faced with the conflicting views of various groups on the environment, economic competitiveness, individual freedom of choice and so on, Mawhinney attempted to facilitate public awareness, dialogue and mutual understanding of the problem. The transport policy process became more accessible to those who had been outside the road policy network. For example, in February 1994 the Greener Motoring Forum was established. Under the joint chairmanship of the Environment and Transport Ministers, representatives of local government, the motor industry, motoring organisations, environmental groups and others with an interest in transport were brought together to make recommendations to the government, as well as to motor manufacturers and others, to promote environmentally responsible motoring. Environmental interests and views were also expressed within the locus of policy debate through the Government's Panel on Sustainable Development and the UK Round Table on Sustainable Development. The Panel, which had direct access to the Prime Minister, was headed by Sir Crispin Tickell, who was a prominent opponent of massive road programmes. Representatives of environmental groups such as FOE and Transport 2000 were included as a chair or a member of transport-related subgroups of the UK Round Table (Dudley and Richardson, 2000: 172–3). The environmental challenge was eroding the traditional road policy at both the cognitive and the institutional levels.

Conclusions that emerged from more than a year of the 'great debate' were documented in April 1996 in a green paper entitled *Transport – The Way Forward*, the first comprehensive statement of policies on inland transport since 1977 (DOT, 1996). The green paper reaffirmed the emerging policy direction away from that based on the 'predict and provide' approach towards 'sustainable mobility'. One of the most notable changes in values was on public transport. In 1990 the environment white paper, analysing the role of public transport, especially rail, in reducing carbon dioxide emissions, had stated that:

> Providing extra [public transport] services would only make emissions worse if few passengers used them: trains are highly fuel efficient when fully loaded, but in practice it is not possible to achieve high loadings all the time.... Even if it were possible to double demand for rail travel by transferring passengers from the

roads, that would only reduce road traffic by about 10%. (Her Majesty's Government, 1990: para. 5.60)

The negative view of the role of public transport remained in the 1994 white paper on sustainable development and the government essentially left to the liberalised markets the task of improving public transport infrastructure (Her Majesty's Government, 1994b: para. 16.36). However, this government attitude was no longer observed in the 1996 green paper, which stated that 'the government believes there needs to be a shift in priorities' to reflect a general public preference, which had emerged from the great debate, for improved public transport over expanded road capacity' (DOT, 1996: 8), and it also promised increased government expenditure on public transport while further cuts were made to the road budget. The green paper, nevertheless, was widely criticised for lacking detail and for shying away from a number of recommendations made by the RCEP. It was reported that, ironically, Mawhinney, in his new role as chairman of the Conservative Party, played a key role in ensuring that the green paper did not contain radical measures that could lead to reduced votes for the Conservatives at the next general election (*ENDS Report* 255, 1996: 20).

The socio-political drive towards the construction of a sustainable transport policy was gathering pace, however. The Road Traffic Reduction Bill, which had been drafted by the FOE and the Green Party in 1993, finally received royal assent in March 1997 after overcoming the final attempt by the motoring organisations to block it in the House of Lords (*ENDS Report* 244, 1995: 34; 264, 1996: 20; 265, 1997: 27–9). The Act required local authorities to formulate measures to reduce traffic. It was, however, heavily watered down during its passage through Parliament and ended up representing more a gesture than a serious government commitment to traffic reductions.

While the transport policy underwent a sea change, policy inertia was also strong. In 1997 the Audit Commission observed that 'In many local authorities the emphasis is still on roads and cars, and relatively few authorities have adopted vigorous restraint measures.... Many of the proposals [for road transport measures] reflect the traditional road-building approach' (cited in RCEP, 1997: paras 7.23, 7.24). This was the same conclusion that the Campaign to Protect Rural England, FOE and Transport 2000

reached (David Davies Associates, 1996). The RCEP also suspected that, in practice, the DOT continued to favour spending on
roads (RCEP, 1997: para. 7.24). At the same time, the government
failed to increase expenditure on public transport. The government's action was 'too little and too slow' and there was 'no sign
as yet of large changes in previous trends' (RCEP, 1997: paras
1.58–85).

Conclusions

This chapter has shown that the British process of integrating
global warming concerns into energy and transport policies was
pluralistic, conflictual and pressure-driven. In the energy sector,
the government was pressured by a broad coalition advocating
energy efficiency and had to grant some small concessions in the
course of the privatisation process. The introduction of the NFFO
was a tactical victory for environmental groups and the renewable
energy industry after intensive lobbying. The pluralist, competitive process of policy integration manifested itself in particular in
the transport sector. The anti-road alliance effectively created a
national political climate within which the authoritative ideas of
the RCEP and SACTRA could have a decisive effect on the policy
discourse. The sway of majoritarian pluralist decision-making
also explains the passage of the private members' bills relating to
energy efficiency and traffic reduction. Flexible coalitions were
successfully built up to win a majority and, once a majority was
formed, policy developed quickly.

Thus, under the institutions characterised by open and competitive policy processes and majoritarianism, the dominant policy
networks were more vulnerable to outside challenges and, in
this sense, were less tight than their Japanese counterparts. The
different political outcomes in road-building policy in Britain
and Japan exemplify this. In both countries the existing road
policy networks came under severe pressure. The Japanese policy
remained fundamentally intact and the British policy underwent
paradigmatic change. In the open, pluralistic political arena,
groups collided and combined dynamically, and this went a long
way to bring about changes in politicians' perception of the
national interest. The changing interests of politicians were well
reflected in an announcement in 1994 by the Labour Party, in

opposition, that it would operate a moratorium on new trunk road schemes if elected (Dudley and Richardson, 2000: 176).

While these are the enabling aspects of majoritarian pluralist institutions, this does not necessarily mean that pluralism has an institutional advantage over corporatism for policy integration. As illustrated by the case of energy efficiency, the policy network advocating competitive prices fought back the challenges on energy efficiency, which set off pluralist political conflict and competition rather than a process of policy integration. This contrasts with the Japanese integration process, in which the political competition between MITI and the EA was more about 'who intervenes'. In Britain, on the other hand, non-intervention had been the dominant policy style on energy efficiency and this was further reinforced by the government's commitment to establishing free energy markets with minimum state interference. The result was political tension over 'whether or not to intervene' in markets.

In the end, the policy network promoting competitive prices, centred on the DTI, the electricity and energy-intensive industries, prevailed. This outcome may not be surprising. After all, the DTI was more powerful than the DOE, and the establishment of free energy markets with minimum state interference had long been a priority for the Conservative government; moreover, the centralisation of political power in the majority party allowed the government adamantly to pursue its commitment to minimum state interference in energy markets. The issue-based approach further elucidates the imbalance of political power between the two coalitions prioritising competitive prices and energy efficiency. State intervention to promote energy efficiency would generate and pass on costs to the electricity producers and major energy-using industries, while its benefits would be dispersed and long term. The result was 'reverse integration': that is, the DTI integrated global warming concerns according to expediency in relation to existing policy priorities.

The British policy style, which rejects state intervention, also posed a constraint on policy integration. The reluctance to take a proactive RD&D policy is a case in point. Nevertheless, this chapter has shown that the government was not wholly passive in the face of this constraint. The Foresight Programme was a new device born out of the government's desire to have a policy akin to a 'developmental' policy without directly involving it in the exercise of 'picking winners'. This did not, of course, suddenly solve the problems associated with the non-interventionist policy style in terms of RD&D; nevertheless, it certainly illustrated that the institutional constraints could be circumvented through policy innovation.

While the policy of minimising state intervention was an important factor in accounting for a weak R&D policy, it was not the sole factor. In the area of renewable energy, the non-interventionist style was compounded by failed attempts in the past to steer technology development, and this reinforced the government's disinclination to establish a large-scale R&D programme. Also crucial was the failure of coordination between macroeconomic policy objectives pursued by the Treasury (reducing public expenditure) and the climate policy objectives. The Treasury's policy objective had, for good or bad, decisive influences on the British policy integration process, as seen in consecutive cuts in expenditure on trunk roads and the failure to increase investment in public transport.

The analysis so far has been more from the perspective of the institutional approach. However, the role of interests that the issue-based approach casts light on should not be marginalised. Coal miners, opposing a massive programme of pit closures, achieved a short-term victory with a further extension of protective measures. The pluralistic conflict over renewable energy was explained above in terms of the asymmetry in political power arising from the logic of collective action between global environmental interests and concentrated local interests. Similarly, in transport policy, local, concentrated interests, which were combined with more diffused environmental interests, were a crucial ingredient in triggering the political dynamic. The issue-based approach, thus, sheds light on the content of pluralist political competition and the outcomes that emerged from it, and, combined with the institutional approach, can help explain British policy integration.

Finally, what conclusion can we draw about the progress of policy integration in Britain? The extent of progress varies from sector to sector and within a sector. Overall, however, the UK Round Table on Sustainable Development and the Government Panel on Sustainable Development expressed disappointment over the progress in the four policy areas I have examined in this chapter, and called for more vigorous intervention and leadership, and strong and strategic policy (UK Round Table on Sustainable Development, 1997: paras 3.12–27, 3.37–49; British Government Panel on Sustainable Development; 1997: paras 20–8). Progress brought about by the political dynamism that was enabled by the majoritarian pluralist institutions was too often offset by the constraints emanating from the same pluralist institutions, combined with the imbalance of political power arising from the structure of interests, and overwhelmed by the policy objectives of the more powerful economic ministries.

Notes

1 The first two NFFO rounds were heavily oversubscribed and, together, were to generate a total of over 624 MW capacity.
2 The ETSU, the House of Lords Select Committee on Science and Technology and the RCEP identified CNG and LPG the most promising alternative fuels, in terms of both their effect on air quality and fuel consumption. See RCEP (1997: para. 2.39).
3 These social groups were not always allied closely against road-building, however. In some more affluent places and during certain periods, established environmentalists and local residents were uncomfortable and more cautious in allying themselves with counterculturalists. See Doherty (1997) and Dudley and Richardson (2000).

9

Interests, institutions and global warming

Since 1988 Japan and Britain have responded to the common threat of global warming. Both countries voluntarily established a policy to tackle the problem before the adoption of the FCCC, and once it was established they developed and implemented policies and measures to meet its requirements as well as the goals they set for themselves. Given that it was a 'framework' Convention, Japan, Britain and all the other signatories were given considerable latitude in designing, developing and implementing policies and measures. The picture that emerges from the present detailed examination of climate policy in Japan and Britain is a mixture of policy similarities and differences. This chapter will bring the findings from the previous chapters together to compare, contrast and analyse Japanese and British global warming policy using the frameworks set out in Chapter 3. The analysis focuses on: the speed of policy change; policy contents, including instruments employed; the degree of policy integration; and policy stringency. The analysis will, then, enable us to go back to, and discuss, the two questions raised in the Introduction: What is the effect of corporatist institutions on a country's ability to tackle challenges to sustainable development? What are the interactions between the institutional and issue-based approaches?

The speed of policy change

The most important change in climate policy was when the countries decided to set national reduction targets for carbon dioxide. This symbolised a commitment to take action against global warming. As predicted in Chapter 3, the consensus-based decision-making style of Japan seems to have made policy change slower than was the case with British institutions, characterised by centralisation of power. In Japan, the EA and MITI had different views on Japan's potential for carbon dioxide savings, based on different data. The EA was representing environmental interests while MITI was representing industry. Decision-making power was shared by the EA and MITI. In the context of conflicting interests, the norm of consensus-building slowed decision-making.

There are two other remarks to be made on the speed of the policy change in Japan. First, it should be noted that MITI accepted the need to have a national stabilisation target only after the release of the Long-Term Energy Outlook in May 1990 and that MITI's stance was based on that Energy Outlook. The Outlook was basically the result of consensus-building between MITI and major energy producers and consumers. This means that the decision on the carbon dioxide target indirectly involved corporatist consensus-building. The negotiation between MITI and the EA was the culmination of long decision-making processes. Secondly, weak political leadership, which exacerbated the slow decision-making, was also at least partly a result of the consensus-based politics in Japan. The tradition of 'consensus articulation' by Prime Ministers and the LDP's factional competition and negotiation for consensus typically result in weak leadership (see Duncan, 2004: 94). In Japan the consensus-based policy style at various levels combined to protract decision-making over the target levels for emissions.

In Britain, on the other hand, the decision-making process was centralised and decisions regarding targets for carbon dioxide emissions were made independently by the government, led by Margaret Thatcher – unlike in Japan, where conflicting interests were brought in and the decision-making process was indirectly a joint one. This allowed Britain to establish and to change its target quickly. Moreover, unlike in Japan, Prime Ministers in Britain are not constrained by the norm of consensus. Prime Minister John Major was instrumental in bringing forward the British target year from 2005 to 2000.

Developments on national targets in Japan and Britain are summarised in Table 9.1. While centralised versus decentralised policy processes appear to have had an effect on the speed of policy change, international influences and the role of individual

Table 9.1 *Development of national targets in Japan and Britain*

	Japan	Britain
Announcement that a national target would be established	June 1990	May 1990
Establishment of a national target	October 1990	May 1990 March 1992: the target year brought forward from 2005 to 2000

June 1992: the Earth Summit

	Japan	Britain
Announcement regarding the target proposal for COP3	August 1997: Prime Minister asks for a decision regarding the target proposal	March 1995: The government calls on developed countries to reduce greenhouse gas emissions by 5–10%
Decision on the target proposal for COP3	November 1997: The target proposal is decided	March 1997: The EU Environmental Council decides burden-sharing targets for member states

politicians were perhaps more important factors. In fact, once Japanese and British policies are put in a comparative context, their importance manifests itself. The similar timing of the first policy change (the establishment of the first carbon dioxide reduction target) cannot be explained without taking into account international influences. The work of the epistemic community on global warming, which pressured, or at least persuaded, the world into taking precautionary action against the threat of global warming, and the following international policy developments, especially the decision to start negotiations on the FCCC at the Second World Climate Conference, prompted governments to change (or establish) policy before the negotiations started. Britain quickly changed its policy stance just after the IPCC report was released (May 1990), while Japan decided (in June 1990) to establish a new policy before the Second World Climate Conference (October to November 1990). The decision to set specific targets in Japan was slow but international political developments created a deadline for the policy.

The effect of international influences was, however, not universal. While a number of countries changed their policies between May and November 1990,[1] some countries, notably the USA, did not change their objections to specific national targets. In both Japan and Britain, what was instrumental in linking international influences and the domestic policy change was the role of an influential politician. In Britain the international demand stemming from new scientific knowledge 'reverberated' (Putnam, 1988) into domestic politics through Margaret Thatcher. Similarly, in Japan it was Noboru Takeshita who delivered the international climate message to the domestic political arena, or more specifically who was instrumental in creating a favourable political climate for the reverberation of international pressure into domestic politics.

Although political leaders, combined with international developments, prompted and accelerated policy changes, as indicated already, their influence was necessarily affected by domestic institutions. In Britain, the change in Thatcher's stance had a decisive influence. In Japan, political leadership is enmeshed in the institutions characterised by consensus-based policy-making. Typically, weak political leadership is the cause and effect of decentralisation of power. The relative success of Takeshita's leadership in climate policy was due not only to his political power but also, and crucially, to his ability to mobilise political support, or a broad political consensus, on the global warming issue.

Nevertheless, in Japan international influences were important stimuli to policy developments. The case of 'pledge and review' during the negotiations over the FCCC was a good example of how perceived international pressure can drive domestic policy developments. The perceived pressure arising from being the host country for COP3 also affected Japanese policy debate over the target proposal for COP3. This sensitivity to international pressure contrasted with British emphasis on national sovereignty, especially in relation to the EU. Lafferty and Meadowcroft (2000b: 424–7) suggest that a foreign-policy orientation towards international cooperation and organisation increases a country's ability to formulate and implement a policy directed at sustainable development. From this perspective, there are three important attributes of Japan that are likely to affect policy. One is Japan's reliance on other countries both for its security (on the USA) and for natural resources (and hence its economic performance). The second is the lingering influence of the early postwar approach to foreign policy, characterised by political passivity in the international sphere accompanied by a focus on national economic reconstruction and growth (Neary, 2002: 162). The third attribute is connected to the second, and actually represents a reaction to it: during the 1990s Japan started to look for a major role in the world (Neary, 2002: 162). This is epitomised by its desire to have permanent membership of the UN Security Council. These attributes seem to have contributed to Japanese sensitivity to international criticism and pressure, which in turn provided an important 'window of opportunity' for domestic developments on climate policy. At the same time, perceived international pressure and international policy developments were often actively used to rationalise and strengthen the different policy stances taken by the EA and MITI.

Britain has very different foreign policy contexts. It is rich in natural resources (though the North Sea reserves of fossil fuels are fast declining) and for a long time was a world power. It has traditionally preferred international action to action at the EU level but has been reluctant to be seen to be influenced by international forces (Weale, 1997: 89–108; Wynne and Simmons, 2001: 105). With these national attributes, international pressure does not have similar influence on domestic politics to that in Japan. Thatcher was influenced by science, which was presented by a leading British scientist. Pressure from the EU over the first British national target, which was set for 2005 instead of 2000, and Britain's isolation among the member states over the EU's carbon/energy tax did not themselves give rise to a policy change.

In summary, the institution-based analysis gives insight into the speed of policy change in the setting of a target for emissions in Japan and Britain; however, the effects of institutions are subject to international influences, policy catalysts and political leadership. Does this analysis indicate there is only a small role for the issue-based approach to explain the speed of policy change? This approach perhaps gives insight into the 'long and wide' consultation exercise employed by the government for the creation of the climate change strategy in Britain. The government's aim was to gain the cooperation of and voluntary commitments from various societal and market actors. Long and wide consultation was a rational means to the desired goal. Policy-makers chose to change policy slowly.

Policy contents and instruments

The main policy contents and instruments that were identified in the previous chapters are summarised in Table 9.2.

Chapters 5 and 7 illustrated a sharp contrast in policy between Japan and Britain. Japanese policy was characterised by a technological approach. With this approach, the government and industry took concerted action, such as joint R&D projects and voluntary action by industry. The provision of a number of financial assistance measures, administrative guidance and a relatively well organised industrial sector facilitated a high degree of policy concertation. MITI provided incentives for desired action but hardly penalised unwanted behaviour. MITI also typically used financial assistance measures in a strategic manner; that is, they were targeted from the long-term view.

British policy, on the other hand, was characterised by a reliance on the workings of markets. British policies were basically

Table 9.2 *Main characteristics of policy contents and instruments in Japan and Britain*

	Japan	Britain
Type of approach	Technological approach	Persuasion and exhortation
Examples	Long-term plans such as New Earth 21 and New Sunshine Programme	The 'Make a Corporate Commitment Campaign'; the Energy Efficiency Best Practice Programme
Strategic characteristics	Strategic financial incentives	Reliance on market mechanisms and limited use of economic instruments
Examples	Financial assistance programme for development and take-up of energy efficiency and 'new energy' technologies	Views on energy efficiency policy; the NFFO; the fuel duty escalator; VAT on domestic energy
Action taken/ preferred	Voluntary action by industry	Lack of strategic intervention
Examples	Keidanren Voluntary Action Plan	Cut in expenditures for renewable energy technologies

aimed at helping 'level the playing field' for carbon dioxide reduction measures and action in markets so as to gain benefits from 'market enablement' effects. In promoting renewable energy, it was the NFFO. In energy efficiency policy, it was largely about informing people. Government intervention was largely limited to the assistance of groups lacking resources so that they could have equal access to built-in economic opportunities. Accordingly, unlike Japanese policy-makers, British policy-makers lacked a strategic interventionist view. The NFFO and the careful avoidance of 'picking winners' contrasted with Japanese targeted investment in photovoltaic technologies. British policy was general and less individualistic in its approach.

Both countries recognised the importance of voluntary action, but the way it was organised was very different. The role of industrial associations was crucial in Japan, as seen in Keidanren's voluntary action, voluntary measures led by MITI and the voluntary purchase of renewable energy by the Electricity Association. In Britain, on the other hand, highly organised collective voluntary action was the exception rather than the rule.

The institutional approach helps explain these contrasts in policy contents and approaches. Japanese policy reflected corporatist and developmentalist policy styles and institutions. Japan had a range of policy tools available for policy-makers to pursue the technological approach and concerted action. Similarly, British policy characteristics reflected its non-interventionist policy style. Collective voluntary action and concerted action were difficult to achieve with essentially pluralist British institutions.

While the institutional approach helps explain the contrasting characters of Japanese and British policy contents, the issue-based approach gives insight into the repeated policy debates over some form of carbon tax and the attempts to introduce one in both countries. In Chapter 3, from the viewpoint of the issue-based approach, I proposed that the configuration of constraints inherent in a given problem limit the range of plausible policies, and rational policy-makers may well end up choosing a similar policy. The problem of global warming is caused by a large number of 'mobile sources' (e.g. people, cars) as well as 'stationary sources' (e.g. power stations, factories), and regulatory intervention, the traditional means of tackling environmental problems, cannot be as readily targeted at the former. Moreover, direct regulation of carbon dioxide emissions could threaten the principle of the liberal market economy. Economic instruments are, then, a rational policy option. This new idea appealed to some policy actors in both Japan and Britain. Nevertheless, neither country could introduce a carbon tax. Indeed, this was an important policy similarity between Japan and Britain. The question of why, however, goes beyond the discussion of policy 'instruments'. A carbon tax or, more generically, the use of economic instruments for environmental purposes is a new policy idea; the advocates of such a tax must first therefore attempt to translate it into actually policy. In this context a carbon tax ceases to have a simply instrumental role *vis-à-vis* a pre-set policy goal of carbon dioxide reduction and in fact becomes a policy goal in itself (Majone, 1989: 116–44). The politics of a carbon tax will, therefore, be fully discussed later.

Policy integration

Chapters 6 and 8 showed that there were sharp differences between Japan and Britain in the pattern of political processes surrounding attempts to integrate global warming considerations into the energy and transport policy areas. Table 9.3 lists the main factors driving and constraining policy integration in the two countries.

Japanese policy integration was controlled, in contrast to more dynamic British processes. In Japan, energy policy-makers, in particular, positively integrated a concern for global warming into their policy-making, and set ambitious targets for energy efficiency and 'new energy', as well as LEVs, and thus sought to 'steer' markets. In this process, however, environmental groups had little role to play *vis-à-vis* the policy networks dominated by industry and bureaucrats. The co-optation and controlled policy integration tended to make the global warming concern a secondary issue to the existing priorities of the energy and transport policy networks.

In Britain, on the other hand, various environmental groups became active in the process of drawing up environment-related energy and transport policies, challenging the traditional policy networks. The process of policy integration was conflictual. The dominant economic interests then fought back against the new challenges. As seen in the case of the promotion of renewable

Table 9.3 *Policy integration in Japan and Britain: driving forces and constraints*

	Japan	Britain
Driving forces	• Co-optation of environmental challenges • Strategic intervention and policy concertation	• Environmental pressures • Centralisation of power enabling ambitious policy changes
Constraints	• Exclusion of environmental interests and actors from policy-making in areas with important implications for global • Co-optation that renders the environmental imperative a secondary issue	• Lack of strategic intervention to steer markets • Pluralist competitive policy processes that make the politics of policy integration prone to be a zero-sum game • Centralisation of power thwarting challenges from environmental groups

energy, there also emerged conflicts between environmental interest groups. The challenges to policy integration resulted in policy shifts to varying degrees. The biggest contrast was between the lack of progress in energy efficiency policy and the change in road policy. For the former, the government clearly prioritised the creation of free energy markets over climate concerns. On the latter, environmental pressure led to the virtual collapse of the traditional (road-building) consensus on which the policy network had been built.

The institutional approach helps explain the different patterns of policy integration. In Japan, under the broad political consensus on tackling global warming, the corporatist policy styles of proactivism and policy concertation between the government and industry enabled and encouraged them to co-opt the imperative of the integration of global warming concerns into their policy-making to a certain extent. Co-optation was a rational strategy to employ in seeking to retain control over policy developments, as it circumvented challenges to the policy core and the fundamental relations of the existing policy network. The other side of the same coin was corporatist exclusion: environmental interests and actors had little access to the core locus of policy-making.

In Britain, on the other hand, the open, competitive pluralist political arena stimulated environmental groups into competing for the government's favour by building a dynamic coalition. As this was combined with a centralisation of political power, drastic policy changes were possible. On the other hand, however, competitive policy processes meant that opposing interests were similarly competing. Centralisation of power could thus also work against environmental interests. In the case of energy efficiency, economic interests won against the emerging energy efficiency advocacy coalition, with the government's uncompromising commitment to non-intervention in energy markets. In addition to these pluralist characteristics of policy processes, another institutional feature, namely lack of institutional capacity for strategic intervention, helps explain the weak long-term investment policy in Britain. The government's Foresight Programme was an innovation designed to overcome this institutional constraint but it can also be seen as a policy for market steering, which was, again, strongly constrained in the British context.

Thus, institutions affected policy integration in Japan and Britain. However, it is misleading to put too much emphasis on institutions in explaining policy integration. In Britain, the government's reluctance to engage in a long-term RD&D policy in relation to renewable sources of energy was at least partly due to past policy failure. More generally, British reluctance to intervene in markets was probably also reflection of Conservative ideology.

Moreover, underlying factors hampering progress in policy integration in both countries often included self-interest. In Japan, the interlocking interests of political actors reinforced the exclusive character of the policy-making process, which resulted in the core issues and conflicts of policy integration hardly being addressed seriously. In Britain, as indicated above, economic interests won over environmental interests on energy efficiency policy in the policy-making institutions characterised by the 'winner takes all' style. A theoretical implication here is that the issue-based approach, which throws light on the importance of interests, needs to be combined with the effects of institutions to explain policy integration. The four previous chapters also indicated that the imbalance of political power between environmental and economic government departments often constrained policy integration. Again on energy efficiency, the DOE lost the battle with the DTI. The EA's failed attempts to introduce a carbon tax are also a case in point. A similar case was seen in Britain when Environment Secretary Chris Patten lost his argument for a carbon tax against the economic ministries, including the Treasury and DTI. Here, the issue-based approach and the logic of collective action implied by it are not entirely irrelevant. Economic policy-makers, who are typically supported by 'concentrated' interests (relatively few, smaller groups with shared aims), tend to be politically more powerful than environmental policy-makers, who are typically supported by 'diffuse' interests (more, larger groups with shared aims). There were, however, victories won by environmental groups. In Japan local protesters and environmental groups got together and successfully put a halt to the construction of a coal-fired power plant. In Britain local residents affected by the construction of new roads were joined by nature conservationists and environmental groups, and their anti-road protests went a long way to bring about a sea change in British road policy. What is important and common here is that these issues involved concentrated interests on the environmental side. The issue-based approach explains a great deal about why environmental groups were successful in these issues.

What conclusions can be drawn from these analyses of the degree of policy integration in Japan and Britain? The emerging picture is very mixed. The degree of policy integration both varied from issue to issue (energy efficiency, transport, etc.) and was influenced by the extent to which environmental actors had been integrated into the policy-making process (on the latter, for example, in Japan policy integration was driven by initiatives by MITI and industry, with little involvement of environmental groups in energy and transport policy processes). Overall, however, policy integration in both countries did not get far in the first ten years of the politics of global warming. Both Japan and Britain failed to realign conflicting policy objectives, and environmental demands were too often a secondary issue. The institutional analysis gave a good insight, especially, into the pattern of policy integration. But institutions alone do not fully account for policy integration. Combining them with the issue-based perspective enhanced their explanatory power.

Policy stringency

Ambition

Policy stringency is about government's ambition in contributing to the world effort to mitigate global warming and also its ability to translate this into a policy and pursue it. Such a policy will necessarily impose costs on society in general and on energy-intensive sectors in particular. The abatement target set for carbon dioxide emissions may give some insight into a government's degree of ambition. However, treating the target as an indicator for policy stringency needs some caution because the opportunity for carbon dioxide reduction will be partly determined by, for example, a country's energy context and population growth. For instance, it can be assumed that if a country has less opportunity to switch fuels, it will experience more difficulty in reducing carbon dioxide emissions. Also, as the Japanese government (especially MITI) repeatedly claimed, countries that have already achieved a high level of energy efficiency will be likely to face difficulties and higher marginal costs in making further improvements in energy efficiency. Modelling studies, indeed, generally show that the marginal costs of carbon dioxide reduction in Japan are higher than those in other OECD countries, although the costs of responding to global warming remain controversial and uncertain, and they are crucially influenced by the types of policy that governments choose (see OECD, 1999: 19–21). In analysing a government's policy ambition, therefore, rather than simply looking superficially at targets or searching for the most credible cost figures, the implications of those targets and the political process of setting them should provide a more useful indicator.

The pre-FCCC targets of Japan and Britain were similar, especially after Britain brought forward the target year for carbon dioxide reduction from 2005 to 2000. The targets can be said to fall somewhere between relatively stringent and moderate. Stabilisation of carbon dioxide emissions was a meaningful first step towards tackling the problem of global warming. Also, targets in both countries more or less met the prevailing international demand for stabilisation of carbon dioxide emissions by 2000.

As with the similar timing of the first policy change, the influence of the prevailing international norm is key in explaining the similarity of the pre-FCCC national targets. Through various pre-FCCC conferences and declarations, stabilisation of carbon

dioxide emissions by 2000 at the 1990 level had become an established norm and a criterion for governments' stance on global
warming. The international norm reverberated in domestic politics
in both Japan and Britain, and again the role played by powerful
policy catalysts, Takeshita and Thatcher, was an important factor
in bringing about convergence in national targets. In both countries, however, the final determinant was economic concerns.

At face value, the target proposals towards 2010 announced
before COP3 differed between Japan and Britain. While the British
target was a 10 per cent reduction in emissions of carbon dioxide,
methane and nitrous oxide, the Japanese target was a 0.5–2.5 per
cent reduction in the same three gases, although with no reduction
being envisaged by MITI for energy-related carbon dioxide emissions. This was far from what international norms would classify
as 'a significant reduction'. A closer examination, however, reveals
that, in fact, the Japanese target proposal was more ambitious
than that of Britain in terms of the actual reductions required. As
MITI claimed, given increases in carbon dioxide emissions since
1990, the Japanese proposal would have to cover the original proposed savings *plus* the increases in carbon dioxide emissions that
had occurred from the 1990 levels. On the other hand, the British
proposal for the target was minimalist, and would involve Britain
in no further measures. Moreover, given projected decreases in
emissions of methane and nitrous oxide with existing measures,
carbon dioxide emissions would even be allowed to increase
slightly from the 1990 levels. It should be noted, however, that the
British proposal assumed that the existing measures would deliver
their expected reductions in GHG emissions fully, and this was the
task that Britain, just as many other countries, including Japan,
had failed to ensure.

International forces combined with individual politicians were
again an important factor in the formulation of the target proposals. Unlike the pre-FCCC targets, however, they had contrary
effects in the two countries. In Japan, under the perceived international pressure that it should exert leadership as the host country
of COP3, Prime Minister Ryutaro Hashimoto supported the EA
in pressuring MITI to be more ambitious. In Britain, on the
other hand, having already achieved a significant reduction in
carbon dioxide emissions, largely from the phenomenon of the
'dash for gas' (see Chapter 7), the international norm allowed

the government to be relatively lax in its efforts to reduce carbon dioxide emissions. In the end, however, in both countries, economic concerns were the crucial factor in shaping the target proposal. In Japan, MITI, which conveyed Keidanren's voice against a stringent target proposal, won the bureaucratic battle with the EA, which wanted a proposal that could live up to Japan's international responsibility as the host country for COP3. In Britain, the target proposed by Environment Secretary John Gummer was an 'antici-pated reaction' to the opposition from economic ministers to a more ambitious target. Despite the apparent differences in the stringency of the two countries' proposed targets, the critical factor that curbed and shaped the target was the overwhelming concern that basic economic policy objectives – employment, international competitiveness and growth – should not be damaged by a highly ambitious policy on carbon emissions.

Lindblom's thesis of 'privileged business' tells us why this con-straint is inevitable (Lindblom, 1977: 70–88). In order to remain in office and, indeed, maintain the liberal democratic system, it is indispensable for policy-makers to maintain the economic welfare of society, but this depends essentially on business performance. There is, therefore, an inherent imperative for governments in liberal democracies to ensure satisfactory business profits. As a result, business is 'privileged' in its participation in and influences on policy-making, and concern about possible adverse effects on business poses 'an all-pervasive constraint' on public policy (Lindblom, 1977: 178).

Once this structural constraint on policy-makers is understood, the implications of the issue characteristics of global warming become apparent. Carbon dioxide is emitted all over the world and its atmospheric concentration does not respect national boundaries, which means that any global environmental benefits that a govern-ment achieves, at some economic cost, are easily free-ridden. Moreover, carbon dioxide emissions are closely linked to economic output and welfare and, unlike most other environmental prob-lems, there is as yet no practicable technological solution available for carbon dioxide removal. The lack of a technological solution, coupled with the ready opportunity for free-riding efforts at any other type of solution, naturally raises the fear that action to abate carbon dioxide emissions will simply damage industrial competi-tiveness and economic performance for little environmental gain.

Not surprisingly, most liberal democratic countries which set pre-FCCC targets at the crest of a global environmental wave made their targets conditional on similar action being taken by their economic competitors, or on future economic conditions.

From this argument it follows that a strong concern for economic consequences in policy-making in Japan and Britain was the corollary of the issue characteristics of global warming, combined with the structural constraints on policy-makers in liberal democracies. That said, the extent to which a stringent policy on global warming would actually cause economic loss was a point of controversy. In Japan, in the process of deciding the target proposal for COP3, MITI projected some dire economic consequences of a 5 per cent reduction in carbon dioxide emissions, while the EA was much more optimistic, arguing that Japan could go further. In Britain, the DOE, in its consultation paper on climate change, revealed that as much as 20 per cent of energy consumption outside the transport sector could be saved at zero cost, or even with net economic benefit (DOE, 1992).

What went hand in hand with the overwhelming concern about the effect on industry and the economy was the bureaucratic power relations between economic and environmental departments. In Japan, MITI had the final say. In Britain, Environment Secretaries Patten and Gummer were, implicitly or explicitly, circumscribed by their position in the departmental pecking order. In the case of Japan, moreover, the bureaucratic power battle had an added dimension in the form of competition over the authority of information and argumentation. As shown, in particular, by the decision regarding the target proposal for COP3, MITI's predominance stemmed, to an important degree, from detailed knowledge and information on energy and industrial matters, which flowed from its close and myriad ties with industry. MITI had infrastructural power, which the EA did not enjoy.

This line of argument, however, raises the question of how some liberal democratic governments were able to set much more ambitious carbon dioxide reduction targets than their economic competitors, including Japan and Britain, since their structural constraints, the issue characteristics of the global warming problem and the weakness of environmental ministries relative to economic ministries are common to all liberal democratic states. One of the more ambitious countries was Germany. In 1990 the former West German government set a 25 per cent reduction target in carbon dioxide emissions from 1987 levels by 2005. This ambitious target was possible as a result of several factors. One was a report by the Enquete Commission on Preventive Measures for Protection of the Earth's Atmosphere, which recommended early action and 'far-reaching reductions' in GHG emissions, later refined to the goal of a 30 per cent reduction in carbon dioxide emissions by 2005 (see Cavender-Bares and Jäger, 2001: 69). The independence of the Enquete Commission, which was composed of politicians and scientific experts, and the consensus within it made the report highly authoritative and successful. Also crucial was the fact that the Environment Ministry won the mandate to prepare a Cabinet decision for a carbon dioxide reduction target. Moreover, the Ministry enjoyed the general support of Chancellor Helmut Kohl, as well as of public opinion. Thus, broad political consensus on a specific target to be aimed at by government, strong political support and a leadership role given to the Environment Ministry in the decision-making process enabled Germany to set an ambitious target. However, a crucial flaw was that the target did not have the support of all government ministries. The Economic Ministry, in particular, did not approve the target or see it as a binding commitment, and it stated that the 25 per cent reduction would be impossible to meet (Cavender-Bares and Jäger, 2001: 72; Schreurs, 2002: 161). The lack of real consensus across ministries resulted in difficulties in developing policy strategies to meet the target (Beuermann and Jäger, 1996: 195–9; Beuermann, 2000: 110).

The economic crisis after German reunification rapidly raised concern about the costs of the climate policy and this resulted in a reduction in its stringency (Michaelowa, 2003: 32). For example, just after reunification the German target was reset at a 25 per cent reduction for the former West Germany and 'significantly more than 25 per cent' for East Germany. The inefficient economic structure of the former East Germany provided an unexpected opportunity for a significant reduction at low (or even no) cost (IPCC, 2001c: para. 8.2.1.2.1). Between 1990 and 1995, carbon dioxide emissions in the united Germany declined by 13 per cent while those in the former West Germany increased by 2 per cent. In 1995 the target was watered down to a 25 per cent reduction from 1990 levels for the whole of Germany (Michaelowa, 2003: 31–2).

In the run-up to COP3, while Germany took the lead in the international negotiations, its target proposal became weaker. The German contribution to the EU's proposed target of a 15 per cent reduction in total EU emissions was a 25 per cent reduction in its emissions of carbon dioxide, methane and nitrous oxide by 2010 (instead of 2005).

From this German experience can be drawn the following analytical conclusions. What enabled Germany to set a more ambitious target were: first, the fact that there existed an institution that was capable of producing authoritative political consensus on a stringent, specific national target which government was recommended to aim at; and second, that the power balance between the Environment Ministry and the Economic Ministry was tilted towards the former, with the official leading role in the target decision-making given to the Environment Ministry and political leadership that favoured the Environment Ministry. However, although these factors made a difference in terms of policy ambition, it has to be recognised that this ambition did not result in specific policy measures, in part because of resistance from the economic ministries, but also in part because economic concerns. In short, over time, the predominant factor shaping German policy was the same as that which curbed and shaped targets in Japan and Britain – overwhelming economic concern.

The politics of carbon tax
As pointed out above, policy stringency is a measure of a government's ability to carry out robust policy, as well as of government's ambition. As a way to analyse government capability in this respect, I will look at the politics of a carbon or energy tax. The reason for the focus on a carbon tax is twofold. First, a carbon tax is widely seen as a rational and necessary policy to tackle carbon dioxide emissions. The IPCC (2001c: para. 8.2.2.1.4) indicates that economic instruments could be the most rational means of implementing climate policy. Barret (1991) also observes that there is 'something close to a consensus that if something substantial is done, then economic instruments will play a role in implementation'. And a carbon tax has been widely considered a key economic instrument (although, since the adoption of the Kyoto Protocol in 1997, which paved the way for international emissions trading, policy-makers have tended to be

more interested in this than in a carbon tax). Indeed, the idea was floated in many liberal democracies as a policy goal in itself. The second reason for examining the politics of a carbon tax is that it encapsulates the fundamental cost–benefit distribution of carbon dioxide abatement policy.

As previous chapters have shown, the case of a carbon tax is an important point of similarity in the first ten years of Japanese and British global warming policy. This similarity was even more significant given the difference in the countries' attitudes to the use of economic instruments for environmental purposes. Britain, which committed itself to the use of economic instruments, including environmental taxes, rejected a carbon tax, just as did Japan, which did not fully accept the idea of environmental taxes. Why were their attitudes towards the use of economic instruments different? Chapter 7 showed that what was crucial in promoting the idea in Britain was the action taken by the Environment Secretary, Chris Patten, and his adviser, David Pearce. Pearce's arguments for the use of economic instruments were especially influential with those government ministries (especially economic ministries) that were reluctant to accept the general idea of environmental taxes (interview with David Pearce, 1998). Patten's support was also crucial. He lent his institutional power to the idea, delivering it to the central locus of policy debate. Patten lost the battle with economic ministers over the introduction of a carbon tax, but managed to obtain government acceptance of the idea of economic instruments or market-based instruments in general. In addition, the market-based approach was ideologically more consonant with the neo-liberal ideology of the Conservative Party than the command-and-control approach (Weale, 1992: 160–1). In Japan, on the other hand, although there were a few experts on environmental taxes, their activities were often confined to work for the EA. Moreover, when the idea emerged, rather than the idea of economic instruments in general, a discrete policy idea of a carbon/energy tax became the focus of discussion and politicians were interested in the revenues from such taxes and gave little attention to their role as a policy instrument for reducing environmental burden. In contrast to Britain, where the general theory of economic instruments was first studied at the ministerial level, in Japan the policy discussion went directly to perhaps the politically most difficult aspect of environmental taxes.

More important in the politics of a carbon tax is how the new idea was processed and given shape. In Britain, the DTI and Treasury, in particular, were greatly concerned about the effect of a carbon/energy tax on the economy and industry, and opposed it. Industry, or more particularly the energy-intensive industrial sector, was successful in winning authoritative acceptance of its claim that a carbon/energy tax would disadvantage competitiveness, despite the lack of empirical evidence for this. In Japan, the early political interest in an environment tax was dampened by strong opposition from business associations, which put pressure on top Liberal Democrats before an election. MITI forecast economic damage – massive unemployment and stringent energy austerity and so on – as a result of the introduction of a carbon/energy tax, represented industry's opposition and foiled any attempts by the EA to pave the way for the introduction of a carbon tax. In so doing, MITI also defended its control of energy-related taxation policy from environmental intrusion. Moreover, close MITI–industry relations enabled both parties to pursue concerted action based on voluntary initiatives orchestrated by MITI or Keidanren, and in doing this they forestalled an environment tax.

Despite the different political processes in the two countries, the outcome was the same: neither introduced a carbon/energy tax. A closer look at the politics of a carbon tax further shows that although the influential national industrial representative bodies, the CBI and Keidanren, strongly opposed the idea of a carbon tax, business did not necessarily speak with one voice; nonetheless, certain voices presided over the policy process. In Japan, it was the steel and electricity industries that were especially influential within Keidanren. In Britain, among the widely different opinions on an EU carbon/energy tax, the interests of the energy-intensive industries and their view of its effects on competitiveness prevailed in the debate. Those who perceived they had most to lose defended their interests most effectively, regardless of the channels through which their interests were mediated.

The issue-based approach throws light on this similar political outcome. Global warming policy is essentially for the benefit of future generations, throughout the world, but that policy threatens carbon-intensive sectors, in particular, with visible and immediate costs. The logic of collective action suggests that a smaller group with a narrow range of interests is more effective in organising itself in pursuit of those interests than a larger group. A large group is often latent, and in order to start political action it first needs to be mobilised (Olson, 1965, 1982). Future generations are perhaps the least capable large, latent group. In this light, the political victory of energy-intensive industries was inevitable.

Of course, the real world is not as simple as suggested. Environmental groups try to act as guardians for future generations but there would also be ancillary present benefits from global warming policy, such as better air quality, less congestion and various gains from improved energy efficiency. Some industries, such as 'new energy' producers and energy service companies, would benefit from a carbon/energy tax and global warming policy more broadly. However, as the IPCC succinctly says, a fundamental problem is that:

> compared to the situation for potential gainers, the potential sectoral losers are easier to identify, and their losses are likely to be more immediate, more concentrated, and more certain. The potential sectoral gainers (apart from the renewables sector and perhaps the natural gas sector) can only expect a small, diffused, and rather uncertain gain, spread over a long period. Indeed many of those who may gain do not exist, being future generations and industries yet to develop. (IPCC, 2001b: 563)

In contrast to the obscure beneficiaries of the policy, the losers have stronger incentives and greater capability to engage in collective action to defend their interests. These factors in turn affect the incentive structure for politicians. The relatively short-term electoral cycle in liberal democracies gives them an incentive to count immediate, concentrated interests, rather than risking office by prioritising long-term, diffuse interests.

Wide public support for a robust global warming policy may change the incentive structure for policy-makers, however, and policy entrepreneurs have an important role to play here by raising public interest in the issue. The problem of global warming policy is, however, that because of its characteristics – such as the complicated cause–effect relations, scientific uncertainty and cumulative effects of the problem – it may not be easy to mobilise *intense* public interest in the issue, and hence pressure and decisive support for policy to tackle it. It should be noted that high environmental awareness does not necessarily equate with strong support for necessary and effective environmental measures.

Moreover, to the extent that global warming policy ultimately entails social, economic and political metamorphosis, the policy would incur costs, whether tangible or intangible, on the public. The costs that a carbon/energy tax would pose on the public are visible and immediate. This may discourage vote-maximising politicians from advocating such a policy. In Japan, for example, most political parties objected to the introduction of an environmental tax in their deliberations on the Basic Environment Law. The British failure to introduce full-rate VAT on domestic energy further suggests that, even if a government with centralised power tries to introduce such a policy, the public opposition emanating from the immediate cost implications can undermine the policy initiative. When the public opposes a policy that imposes costs on them for the benefits of future generations, the interests of the former will prevail within liberal democratic political processes.

What these arguments about the politics of a carbon/energy tax suggest is that those who perceive concentrated losses have an inherent political advantage in defending their interests, because the issue characteristics bring about a serious problem of collective action for policy beneficiaries; moreover, the incentive structure for politicians that underpins any democratic system prompts them to favour the potential losers in any policy innovation, whose political voice is generally louder than that of those who stand to gain from it. And the similar political outcome in Japan and Britain indicates that this is so regardless of institutional characteristics.

The implication of this line of argument is profound. If it is the built-in character of the issue that is the deciding factor for the fate of the policy, other liberal democracies are likely to share the similar traits in policy. Is this claim sustained? A quick look at policy developments in other countries throws this claim into doubt. Starting with the introduction of a carbon dioxide tax in Finland in 1990, taxes on emissions had been introduced in most Nordic countries by 1997. One of the crucial ingredients for the successful introduction of a carbon/energy tax in these countries was political leadership at the peak of environmental awareness in the late 1980s and the early 1990s. Major political parties and governments in these countries kept up with their environmentally focused parties in public policy documents, such as party programmes and white papers (Midttun and Hagen, 1997). Nevertheless, the carbon/energy taxes introduced in these countries reveal strong constraints from industrial opposition. Energy-intensive industries, in particular, won tax exemption or pay only a 'symbolic' rate, despite the fact that such loopholes undermine the rationality and efficiency of the policy (OECD, 1995: 111; Midttun and Hagen, 1997; IPCC, 2001b: para. 8.2.2.2).

In Norway, for example, a carbon dioxide tax was introduced in 1991 as part of a major tax reform, as well as a general trend towards more use of economic instruments in environmental policy (Reitan, 1998: 16). Despite opposition from industry and some political parties, the minority government managed to gain majority support in parliament for a relatively high carbon dioxide tax. The successful introduction of the tax highlights the environmental ambition of Norwegian politicians in general. Nevertheless, extensive exemption and reduction schemes were incorporated with a view to protecting the competitiveness of energy-intensive industries, while tax deduction rules reduced the real tax rate on offshore oil companies to only about a third of the nominal rate (Reitan, 1998: 17; Kasa, 1999: 9). The government's attempts to broaden the tax base, increase the tax rate and eliminate exemption were hampered by opposition from energy-intensive industries and their workers – which dominated the peak industrial association and the trade union, respectively – through corporatist channels and links with major parties (Reitan, 1998; Kasa, 1999; Langhelle, 2000: 186–94). Between 1991 and 1993, the taxes for industry were gradually lowered while those on petrol for consumers were raised (Reitan, 1998: 17). After a decade, a proposal to tax energy-intensive industries finally appeared on the parliamentary agenda. The coalition government could virtually shut the corporatist channels for industrial pressure in the policy-making process. However, the energy-intensive industries, together with the unions, campaigned effectively, focusing on influential politicians of major opposition parties and those members of Congress who represented regions with energy-intensive industries, and succeeded in scrapping the minority government's proposal (Kasa, 1999: 18). Thus, the energy-intensive industries won 'a substantial victory' in the politics of a carbon dioxide tax (Kasa, 1999: 20).

In Japan, the EA's attempt to make headway in introducing an environment tax was thwarted by the more powerful MITI, which represented industry's interests. In Britain, key political leaders were too concerned about the impact of a carbon/energy tax on international competitiveness. The Norwegian case, however, suggests that even if politicians have the will to introduce an ambitious policy, those who stand to lose can take effective collective action to defend their interests through a variety of political channels over a period of time.

The issue-based approach explains a large part of the stringency of global warming policy. The issue characteristics gave rise to considerable concern, on the part of government but more especially the powerful economic ministries, about the effect an ambitious policy would have on the economy and industry. However, the fundamental constraint stemmed from the asymmetry in political power arising from the configuration of interests involved in the issue. Under the liberal democratic system, the imbalance of political power arising from the logic of collective action is critical between energy-intensive industries (on which costs fall) and future generations (who are the primary beneficiaries). Britain's failure to introduce the full rate of VAT on domestic energy further suggests that the most affected group, even if it has dispersed interests, is still capable of undermining a government policy that is directed at the benefit of future generations – even a government with institutions characterised by centralisation of power and little scope for the use of a veto against it. The political ability to introduce stringent global warming policy is severely undermined by the constraints intrinsic to the issue.

Interests, institutions and global warming

Despite the relative similarity of Japanese and British responses at the end of 1980s and in the early 1990s, over time their positions at the international negotiations diverged. At home, the policies they developed to meet their international obligations differed in both content and strategy. Patterns of policy integration, which indicates qualitative progress on global warming policy, were also different. At the same time, their capacity to implement robust policies to take their due part in mitigating global warming was often constrained in a similar manner. I have shown that consensus corporatism and majoritarian pluralism, as contrasting institutional systems, give a good vantage point for analysing global warming policy in Japan and Britain. However, this institutional approach has little explanatory power in accounting for the fundamental constraints on carbon dioxide mitigation policy. For this, we have to turn to the characteristics of the global warming problem and their effects on the political behaviour of actors operating in liberal democracies. International influences, crucially mediated by influential politicians, also had converging effects on policy.

I have analysed this mix of policy differences and similarities by discretely examining four aspects of policy. Overall, however, the policy that Japan and Britain developed to meet their commitments to reducing carbon dioxide emissions can be seen as an outcome. This outcome of policy-makers' attempts to tackle the problem (understood as a configuration of constraints) has a structure: a core surrounded by peripheries. What emerges is the centrifugal nature of policy. The most affected actors resist robust policy, undermining the policy core; in response, various policy measures are developed surrounding the policy core, often in a way that avoids addressing the crucial issue directly. These peripheral policy measures are nationally distinctive, reflecting institutional opportunities and constraints, the main discourses in the relevant policy domains and existing policies.

This analysis of global warming gives insight into one of the questions that I raised at the beginning of this book. Relative to a more pluralist society, does corporatism make it easier or harder to take on the challenges to sustainable development? The question needs to be answered carefully. The study is only a two-country comparison and 'corporatism' here has essentially focused on government–industry relations. Nevertheless, taking into account the experiences of Germany and Norway, which are widely seen as environmental leaders and as having consensus corporatist institutions, the study can suggest that corporatism does give rise to some policy differences which may be environmentally beneficial; however, it does not make a decisive difference to the national capacity to implement a policy that is crucial for the transition to sustainable development – the capacity to overcome the fundamental constraints arising from an issue that concerns the interests of future generations.

Institutions and issue characteristics
Another question I raised in this book concerns the relationships between the institutional and issue-based approaches. Although I have argued that the policy core is better explained from the

perspective of the issue-based approach, I do not intend to place institutions secondary to issue characteristics in policy analysis. Nor did this study negate institutional effects on political motivation, behaviour and outcomes. The question is not which of these approaches is right, but concerns rather the relationships between the two. What interactions have we discerned? How is their explanatory power enhanced by combining the two? This study has shown three patterns of interaction between the two approaches.

(1) *Issue characteristics (understood as a configuration of constraints) limit the range of practical and rational policy measures and instruments from which institutional selection takes place.* We have seen that institutions affect governments' capacities and preferences for certain policy approaches and instruments. Japanese corporatist and interventionist institutions enabled Japan to have a proactive technological approach and a well organised voluntary approach. On the other hand, the British non-interventionist, indirect approach reflected the lack of such institutions in Britain. This analysis, however, does not necessarily rule out the effect of issue characteristics. As the idea of the use of economic instruments was repeatedly floated in both Japan and Britain, the characteristics of the global warming problem broadly determine what are rational policy approaches.

(2) *Issue characteristics, interacting with institutional characteristics, affect how institutions matter for political outcomes.* Institutions have two faces. They enable as well as constrain. Issues, by presenting a particular configuration of costs and benefits to political actors and so partly determining their incentive structures, condition the institutional effects on political behaviour and outcomes. The consensus corporatist institutions of Japan enable it to take proactive, concerted action towards a long-term goal which may not necessarily be rational to pursue in the short term. We are likely to see these enabling effects of institutions when the cost–benefit configuration of an issue, as perceived by the affected actors, embraces opportunities for positive-sum or at least zero-sum solutions. Such an issue, combined with corporatist institutions, gives actors incentives to pursue the strategy of policy concertation. However, when an issue is perceived to present

opportunities only for losses and no positive-sum solution, and to challenge the interests, values and relationships embedded in the corporatist institutions, the institutions would have the opposite effect on policy. The corporatist actors would then have an incentive to defend their institutionalised interests and relationships and the exclusive nature of corporatist policy-making would enable them to resist the new policy. Policy developments on such issues would be likely to be paralysed and the policy stifled. The contrasting policy developments in relation to 'new energy' and LEVs, on the one hand, and an environment tax and road-building give examples of how issue characteristics affect institutional effects on policy outcomes. While in the former, MITI and industry saw new market opportunities, and thus had an incentive to take concerted action, the latter cases were simply perceived as 'loss imposition' or threats to the institutionalised interests and relationships, thus resulting in policy stagnation.

The majoritarian pluralist institutions of Britain also present opportunities and risks. The institutions characterised by fewer possibilities of veto and the 'winner takes all' decision-making style enable Britain to make quick, innovative policy changes. This institutional capability is more likely to be realised when an issue has the capacity to mobilise intense public attention, when the new policy or policy change gives rise to visible and immediate beneficiaries, and when the losers (if any) are obscure. This is because the 'winner takes all' decision-making style makes which group or coalition wins any particular political competition a factor that determines the outcome; the affected groups' incentives and capabilities to engage in collective action and coalition-building are therefore crucial. Once they have won the government's favour, majoritarian institutions will increase a country's capacity for innovation.

When the costs are visible and immediate but the benefits are obscure, the losers will have greater incentives to engage in collective action. These issues are likely to show the constraining effects of institutions. Policy development can be hampered by an opposition coalition, or even if policy does change, the policy will be vulnerable to political challenges from losers, creating the risk of a policy U-turn or simply a failure to be implemented. The sea change in policy on road-building and the failure to introduce the full rate of VAT on domestic power and energy are examples of how different issue characteristics lead to opposing institutional effects on political outcomes. On road policy, there were concentrated beneficiaries (local residents) in policy change, and the visibility and particularity of the damage road-building does (e.g. to the landscape) attracted intense public focus, which in turn contributed to the building up of a broad alliance in support of policy change. On the other hand, VAT on domestic fuel and power generated visible and immediate costs, especially on the 'fuel poor', while no definable social groups benefited from it. The planned increase in the rate of VAT imposed was effectively challenged by the losers and the government failed to implement it.

Although the argument here stresses the role of issue characteristics in determining how institutions matter to the political outcome, they are not the sole factor influencing institutional effects. Political conditions and individual actors, in particular, are also crucial factors (Weaver and Rockman, 1993). For example, lack of party cohesion would undermine the decisiveness of majoritarian institutions. Also, it is political actors' perception of the distributional effects of an issue that matters in how issue characteristics determine the institutional effects on political outcomes. Actors may have different perceptions of the same issue, according to their institutional roles and relationships, past experiences and new information and knowledge, as seen in different cost–benefit analyses of global warming policy between the EA and MITI in Japan. I do not discuss this question further here, but will come back to this point after discussing the third interaction between issue characteristics and institutions which this study has observed.

(3) *Issue characteristics override distinctive institutional effects on political outcome.* One of the implications of the second interaction explained above is that when consensus corporatism and majoritarian pluralism have constraining effects on political outcome, they can still lead to similar political outcomes. This is because both are resistant to policy developments on certain issues. In other words, over certain issues institutional effects lose their importance in explaining political outcomes and then issue characteristics become the deciding factor.

This book has shown that this is likely to happen when an issue generates an overwhelming imbalance of political power between affected groups. In the case of a carbon/energy tax, its principal distribution of costs and benefits gave rise to a political competition between future generations throughout the world and energy-intensive industries. While the future benefits were obscure, the energy-intensive industries perceived immediate costs for them. Not only was the gap in the capacity for collective action between the two parties obvious, but also, under the present democratic institutions, self-interested politicians had incentives to favour those bearing the costs ('the losers'). With this kind of issue, regardless of whether it is processed through consensus corporatist or majoritarian pluralist institutions, losers would have decisive influences on policy.

Although I have examined the three interactions separately, they are not mutually exclusive. To the extent that the choice of policy instrument is often the policy goal and the object of political contention, the first pattern of interaction is related to the other two patterns. Also, the third pattern of interaction is, in a sense, an extreme case of the second one.

It should also be noted that issue characteristics are not *the* factor determining how institutions matter to political outcomes. Weaver and Rockman (1993), for example, argue in their book *Do Institutions Matter?* that institutions' effects are contingent on political and social conditions, other institutions and existing policies. Global warming policy in Japan and Britain indeed confirms their claim. Electoral vulnerability undermined institutional capabilities to make important changes to transport policy in Britain. On the introduction of the full rate of VAT on domestic power and energy, weakened party cohesion heavily undermined the institutional opportunities. In Japan taxing coal was even more difficult because of the linkages with other policies.

After all, it is through political actors' intentions that institutions have their effects on political outcomes. Actors define the cost–benefit configuration of an issue, which, in turn, affects how institutions matter to political outcomes. With consensus corporatism, where there is congruity between the cost–benefit implications of an issue and the actors' interests, the actors' cognitive capacities to find win–win opportunities are particularly important. Japanese global warming policy also indicated that actors can strategically use institutional opportunities to defend their interests. Policy concertation, seen in various voluntary actions by industry, was a case in point. With majoritarian institutions, the intention and beliefs of the governing party and political leaders were crucial. In the road-building policy, it was through the Chancellor's opportunistic behaviour and the Transport Secretary's intentions to embark on policy integration that the British majoritarian institutions had enabling effects. This case also showed the importance of the cognitive dimension of policy processes. The new, authoritative ideas about latent traffic demand certainly affected political actors' perception of the costs and benefits associated with road-building. In Chapter 3 I suggested that the institutional approach gives scope for ideas to alter the definition of a problem and interests, while the issue-based approach sees ideas instrumentally. Both views have validity.

Without considering institutional incentives and conditioning, the issue-based analysis leads nowhere. Without looking at issue characteristics, the institutional analysis can be misleading. They complement each other and by looking at their interactions we are better positioned to explain policy. This book has shown examples of patterns of interaction between the two. At the same time, policy is a product of dynamic processes in which various endogenous and exogenous factors come into play, affecting the interactions. These factors must themselves be understood.

Conclusions

This book has compared and contrasted global warming policy in Japan and Britain, countries seen as having important institutional features of consensus corporatism and majoritarian pluralism, respectively. These institutional characteristics helped explain global warming policy in Japan and Britain. Nevertheless, what stood out in the analysis of policy was the importance of issue characteristics. The nature of the problem has crucial effects in shaping actors' incentive structures, political behaviour and, hence, policy. Tackling global warming is peculiar in its character as an ultimate public good, which transcends time and space in the human world. Any sacrifice made to this end is diffused accordingly. The sacrifice can be easily free-ridden, while producing little or no difference to the global climate. Rational actors on whom concentrated costs fall would not only resist but also suspect the legitimacy of such a policy. Without a popular mandate, combined with the office-seeking behaviour of rational politicians, it would not be easy for a country to implement substantial measures. In the absence of strong political leadership, global warming politics will easily become subject to bureaucratic competition, in which, typically, a weaker environmental department has to confront a stronger economic ministry. Global warming is a difficult problem to solve politically as well as technically. Certainly the political leaders in many liberal democracies were too optimistic about their capacity to control and cut carbon dioxide emissions when they established national reduction targets in the early 1990s, a period of international environmental zeal. In the first several years of global warming politics, policy-makers learnt how hard their self-imposed challenge was and that politically difficult, more stringent measures would be required to pursue the ultimate goal of the FCCC.

Although this study took Japan and Britain as examples of consensus corporatism and majoritarian pluralism, respectively, I do not intend to give the impression that their policies are typical of these institutional characters. The economic and energy context, policy legacies and detailed policy-making processes differ from country to country and this will affect the contents of policy. Also, there are variations in consensus corporatism and majoritarian pluralism. In particular, 'corporatism' has other important characteristics which Japan does not necessarily share. For example, if Japanese institutions are seen as having the characteristics of 'exclusive' corporatism, it is 'inclusive' corporatism, as may be found in the Netherlands, Sweden, Denmark, Norway and perhaps Germany, that is widely viewed as characterising the environmental leaders (Lafferty and Meadowcroft, 2000b; Andersen and Liefferink, 1997).

That said, there is room for at least some preliminary thoughts on which of the contrasts between Japanese and British policies are likely to be generalisable and which factors are unique to Japan and Britain. Interventionism, strategic orientation and market steering, which are associated with corporatist institutions, are likely to generate proactive policy responses, while countries with a tradition of non-intervention are likely to respond reactively. For example, Japan and other countries with corporatist characters such as the Nordic countries generally employed a wide range of policy instruments, from regulation, incentives, R&D to voluntary action. On the other hand, Britain, the USA, Australia and Canada, which have pluralist institutional characteristics, tended to rely largely on voluntary commitments from industry. Unlike corporatist countries like Japan, Germany and the Netherlands, such voluntary commitments in pluralist countries received little support in the form of other measures to buttress their effectiveness (Lafferty and Meadowcroft, 2000b: 398). Interventionist and close government–industry relations in corporatism made various policy instruments readily available to governments. In the USA, Australia and Canada, in particular, such additional measures to support voluntary commitments tended to come or be proposed in response to the realisation that voluntary commitments alone would be inadequate (see Gay *et al.*, 1999: 10; Zhou *et al.*, 1999: 11; Nyenzi *et al.*, 1999: 12).

Japanese and British policy on RD&D relating to climate is also likely to be more generally applicable to other countries. Pluralist, non-interventionist countries including Britain tend to have a weak long-term RD&D and commercialisation policy. In contrast, strategic intervention and active state steering were important factors that have made Denmark the world's largest producer of wind turbines and Japan the world's largest producer of solar cells.

Corporatism, however, will not always have positive effects on environmental policy. In the same way that the politically influential Keidanren protected those who would lose from the imposition of a carbon tax, peak associations in corporatism may tend to represent those members who will lose rather than gain from a policy. In Norway and Denmark, for example, industrial peak associations represented losers from a proposed carbon tax, even though it would have benefited some of their other members (Kasa, 1999: 14–15; PETRAS project, 2002). Moreover, even where there is a more inclusive form of corporatism, as in Sweden, Denmark and Germany, environmental groups tend to be marginalised or excluded as negotiation proceeds on the details of policy (see, for example, Kronsell, 1997: 57; Pehle, 1997: 178; Svendsen, 2001: 28).

Nevertheless, the role of environmental groups in decision-making is an important factor that distinguishes Japanese environmental policy-making from that in inclusive corporatist countries. In Japan there remain the effects of the systematic exclusion of environmental groups. Moreover, the EA's institutional competence over environmental policy implementation was very limited, and when it came to climate policy the whole energy issue was under MITI's jurisdiction. MITI, therefore, had a key role in policy-making and, with its overwhelming power, marginalised the EA and its loosely allied environmental groups in processing important issues relating to climate policy. As a result, in contrast to the 'environmental' corporatism (involving an environmental ministry, industry and environmental groups) in European countries, what often featured in Japanese climate policy and especially the associated implementation processes was the predominance of 'economic' corporatism, involving MITI and its close industrial allies (which tend to affiliate themselves to a single ministry). Still, the EA had competence over overall global warming policy. Having different 'philosophies', the EA and MITI were the locus of state pluralism, while at the same time implementation was characterised by corporatist policy concertation. These institutional characteristics are specific to Japan and are important factors that gave rise to, *inter alia*, fierce, pre-emptive conflict between the EA and MITI over target-setting and policy integration.

Another important factor specific to Japan is the role of the LDP, which dominated Japanese politics from 1955, albeit to a lesser extent after 1994. During its long reign a productivist political framework was built up, based on close relationships between the LDP, big business and the economic bureaucracy, from which environmental interests, after being co-opted in the early 1970s, were systematically excluded. A related issue here is the electoral system, which tended to favour the LDP. Although the multi-member constituency system up to 1995 encouraged small parties, it generated low levels of electoral competition (McCargo, 2004: 118). Moreover, the electoral campaigns, largely based on provision of particular goods rather than party platforms, marginalised environmental issues. Furthermore, strong factionalism within the LDP and the norm of consensus-based decision-making put constraints on strong prime ministerial leadership. Taken together, these factors have weakened political support for the environment, whereas in some European countries an electoral system that features proportional representation and consensus-based legislative decision-making have given a boost to environmental interests.

Turning to Britain, there are at least two important factors specific to British policy processes. One is the effect of the EU. EU policy on climate change may have made British policy less reactive and richer than otherwise would have been the case; examples are EU-wide energy labelling and minimum standards, the requirement to introduce national programmes for energy efficiency, and funding for energy efficiency and renewable energy. The role of the EU in the climate politics has been expanding, especially since 1997. At the same time, Britain is becoming more strategic in dealing with EU policy development, by trying to 'upload' its own policies rather than 'downloading' EU policies inspired by other member states, many of which have corporatist institutional characters (Jordan, 2002). An implication here is that Britain may become more active in environmental policy-making than other pluralist countries and that, if not, British policy may be strongly influenced by a more interventionist (or corporatist) environmental policy of Europe's leading environmental states.

Another important factor that distinguishes Britain especially from the USA is the centralisation of power. The winning party has almost unhampered power, with little in the way of veto points in policy-making. In this context, the role of the Treasury, with its overwhelming power, is of particular importance in integrating environmental policy into economic policy or rejecting it. The radical policy change as seen in the process of policy integration in the transport sector is possible with the high institutional capacities for innovative policy change.

This discussion about the generalisability of Japanese and British policy responses raises further research questions rather than neatly concluding the study. How are these generalisations valid? Are there any analytically useful subcategories of consensus corporatism and majoritarian pluralism? What are the effects of a general tendency to broad participatory (rather than competitive) policy processes on environmental policies under consensus corporatism and especially majoritarian pluralism? Research should also be conducted across different environmental policy issues to understand better the relationships between institutions and issue characteristics.

These questions may indicate only a preliminary contribution of this work; nevertheless, I hope this study has thrown new light on the analysis of environmental policy and more broadly sustainable development, while at the same time contributing to the institutional approach by addressing the importance of issue characteristics and the interaction between them in policy analysis.

Note

1 According to Rowlands (1995: 79), between May 1990 and December 1990 fourteen of the OECD's twenty-four member states set GHG control targets.

10
Epilogue: after the Kyoto conference

The adoption of the Kyoto Protocol in December 1997 brought global warming politics to the next stage. With legally binding specific GHG reduction targets allocated to developed countries and as many of them had failed to control their emissions of carbon dioxide, efforts at reducing GHG emissions were strengthened in many developed liberal democracies. Japan, which failed to achieve its 2000 target, and Britain, which was one of the few countries to meet its targets, are not exceptions. I will briefly examine major policy developments in the two countries and discuss the implications for the main findings in this book. The change of government in Britain in May 1997, when the Labour Party won the general election with a landslide after eighteen years of Conservative government, also provides an interesting test of whether and to what degree British policy characteristics are the consequences of deep-seated state orientations underpinning its institutions or specific party ideologies.

Policy developments in Japan

Climate politics and policy in Japan after COP3 have centred on the Global Warming Prevention Headquarters in the Cabinet headed by the Prime Minister and the Joint Meeting of Councils Relating to Domestic Measures to Arrest Global Warming, which was first convened by the then Prime Minister, Ryutaro Hashimoto, in August 1997, and which has brought together key individuals from the relevant advisory councils with the aim of facilitating policy coordination among ministries.

The main policy frameworks after the Kyoto conference were the Guidelines for Measures to Prevent Global Warming (henceforth simply Climate Policy Programme) and the Law Concerning the Promotion of Measures to Cope with Global Warming (henceforth Global Warming Law), both of which were adopted in 1998. The Climate Policy Programme was decided by the Global Warming Prevention Headquarters in June 1998 as the strategy to achieve a 6 per cent reduction in GHG emissions by 2008–12, as required by the Kyoto Protocol. It was based on the report made by the Joint Meeting of Councils in November 1997 (Joint Meeting of Councils Relating to Domestic Measures to Arrest Global Warming, 1997), and this report was in turn largely underpinned by the outcomes of the separate deliberations on climate policy by MITI and the transport-related ministries (the MOT, the MOC and the National Police Agency). The Climate Policy Programme is composed mainly of MITI's policy to strengthen the Energy Conservation Law and measures set out in the Long-Term Energy Outlook drawn up by MITI's Advisory Committee for Energy. The Global Warming Law, on the other hand, is based on pre-Kyoto deliberations by the EA. This was the world's first law dedicated specifically to the global warming problem. However, in contrast to the strategic and programmatic nature of the Climate Policy Programme, the Law is essentially a basic law, in that it provides an integrating framework for implementing global warming policy in general and not a strategy to attain the Kyoto target. Both the Climate Policy Programme and the Global Warming Law were revised when the government ratified the Kyoto Protocol in 2002 to strengthen policy in view of rising GHG emissions.

There was much criticism of the Climate Policy Programme from environmental groups, centring on the lack of transparency and participation in its production, as well as its contents. One of the controversies raised was the fact that rather than the Law laying down the Policy Programme, the Programme – decided on by Cabinet – lays down the Law. This problem was highlighted when the Central Environment Council deliberated on the Basic Policies Relating to Global Warming to be established under the Law.

Controversy also surrounded the breakdown of the 6 per cent target stipulated under the Climate Policy Programme. The subsidiary targets were based on an agreement struck between the EA and MITI just after the Kyoto conference. Simply put, the government maintained the official pre-Kyoto target of a 2.5 per cent reduction in carbon dioxide, nitrous oxide and methane by 2008–12 from 1990 levels, with a 0 per cent reduction in carbon dioxide emissions from energy-related sources. The remaining 3.5 per cent (of the 6 per cent target) was to come from carbon dioxide removal from sink activities, such as afforestation, reforestation and the 'Kyoto mechanisms' (see Chapter 2). There was much uncertainty about whether Japan would receive, as the government wanted, the approval of the international community for its reliance on sink activities to meet the target; however, the US withdrawal from the Kyoto Protocol in 2001 and the EU's determination to keep the Protocol alive gave Japan an unprecedented diplomatic advantage, and the government did in fact gain such approval, for up to 3.9 per cent of the target to be met through sink activities. Given that the original estimate of carbon dioxide removal from forestry and other sinks in Japan in 2010 was about 3.7 per cent of total carbon dioxide emissions at 1990 levels, the 3.9 per cent credit was an extraordinary diplomatic success for Japan. Details of the current subsidiary targets are shown in Table 10.1.

The competition between the EA and MITI regarding which would have jurisdiction over climate policy in the industrial sector continued after COP3. Their rivalry became heated when the EA made repeated moves to introduce regulatory measures for carbon dioxide emissions from the industrial sector in the Global Warming Law. MITI and industry severely criticised the move, claiming that it would double-regulate industry since MITI's Energy Conservation Law had already functioned as a regulatory measure (if a soft one) over industry's carbon dioxide emissions. The eventual retreat by the EA indicates MITI's victory in this jurisdictional competition. It should be noted, however, that Keidanren also came out as a major winner. Keidanren conceded the claim that its Voluntary Action Plan on the Environment lacked transparency by agreeing that progress in implementing the sectoral action plans could be examined by relevant advisory councils; however, it succeeded in making the Action Plan the main energy conservation measure, while staving off government interference but still gaining financial and other support to increase the Action Plan's effectiveness.[1]

Table 10.1 *Japan's subsidiary targets in the 2002 Climate Policy Programme*

Target	Percentage change from 1990 levels
Energy-related emissions of carbon dioxide	0
Industry	–7
Residential and commercial	–2
Transport	+17
Non-energy-related emissions of carbon dioxide, methane and nitrous oxide	–0.5
Innovative technological development and further promotion of activities to prevent global warming undertaken by Japanese people from all sectors of society	–2
Technological development	–0.6
Voluntary activities by individuals	–1.3 (estimated maximum –1.8)
Hydrofluorocarbons, perfluorocarbons, sulphur hexafluoride	+2
Removals by sinks (e.g. forestry)	–3.9
Kyoto mechanisms, etc.	–1.6

The 2002 Climate Policy Programme took a step-by-step approach; assessment and review of measures were scheduled for 2004 and 2007. The Programme focused on the 'first step' period (2002–4) and it largely relied on Keidanren's Voluntary Action Plan, which accounted for a third of carbon dioxide savings from energy conservation policy. The EA's original vision was to introduce, in the 'second step' period (2005–7), new measures, such as an obligation on factories and businesses to monitor and make public their GHG emissions, negotiated agreements replacing Keidanren's Voluntary Action Plan, domestic emissions trading and what the EA called a global warming prevention tax. (There was also a 'third step' period, to take the programme from 2008–12.) However, the EA failed to implement this vision: industry, represented by Keidanren, successfully opposed the new measures and, consequently, the thrust of the Climate Policy

Programme became instead the full implementation and strengthening of existing measures.

This does not mean, however, that there is nothing new about Japanese climate policy after COP3. One of the innovations was MITI's 'top-runner approach' to setting standards for vehicles and electrical appliances. Instead of setting a lower limit for the average efficiency of each manufacturer's products, standards were now to be established at levels which met the highest energy efficiency achieved among products currently available. It changed the concept of 'standards'. They were no longer set in a bid to eliminate the least efficient, but set to drive technological innovation.

The US withdrawal from the Kyoto Protocol in March 2001 was the biggest setback in climate politics in Japan after COP3. Keidanren and the Chamber of Commerce and Industry in particular were strongly opposed to Japan's ratification of the Kyoto Protocol, and hence the implementation of climate policy, without the USA's involvement. Without industry's cooperation, the government's Climate Policy Programme would be jeopardised (see Table 10.1). It was reported that when the Marrakech Accords were agreed at COP7 in November 2001, as a gesture of courtesy, Prime Minister Junichiro Koizumi personally contacted Takashi Imai, chairman both of Keidanren and of Nippon Steel, about the government's intention to move towards ratification (*Yomiuri Shimbun*, 21 November 2001). However, even after the announcement of the government's intention to ratify the Protocol, opposition remained, especially in the steel and electricity industries, and, coupled with sceptics within the LDP and one of its coalition partners, the Conservative Party, the submission to the Diet of the proposal to ratify the Protocol and the bill to revise the Global Warming Law was delayed (*Asahi Shimbun*, 8 April 2002).

While industry, especially Keidanren, has had a strong sway on the government's climate policy, environmental groups have been rapidly growing in strength since Kyoto. The representative of the Kiko Network, Mie Asaoka, now regularly attends the Central Environment Council. MITI's Advisory Council for Energy also, for the first time, included, in its subcommittee, three members representing environmental groups and a group against nuclear energy. Moreover, they have not only challenged the effectiveness of the government's policy on carbon dioxide emissions (see, for example, Tsuchiya, 2001), but also started to question the basic picture of carbon dioxide emissions that underlies that policy. The government, or more specifically MITI, emphasised that carbon dioxide emissions from industry were declining and that the main sources of rising emissions were the residential/commercial and transport sectors. MITI and industry also argued that industry was already fulfilling its responsibility through, *inter alia*, Keidanren's Voluntary Action Plan. The main policy implication was that policies and measures should be targeted at the household and transport sectors, with a special emphasis on the need to change lifestyles. MITI also argued that Japan's good record on energy efficiency in terms of GDP owed much to significant improvements in the use of energy in the industrial sector after the oil crises (this claim became established discourse in the Japanese climate policy debate). Environmental groups recognise the importance of raising awareness and changing lifestyles; however, they point out that half the emissions from the residential/commercial and transport sectors have sources related to business activities, which Keidanren does not cover in its Voluntary Action Plan, and that high energy efficiency in Japan in terms of GDP in large part derives from the relatively small energy consumption of the residential/commercial and transport sectors.

The EA brought these arguments to the deliberations over the new Climate Policy Programme to be established in 2002. The EA pointed out that about 70 per cent of carbon dioxide emissions were from activities related to industrial operation and, counter to MITI and industry's claim that industrial efforts based on Keidanren's Voluntary Action Plan had been effective, it argued that carbon dioxide reductions in the industrial sector largely arose through improvements in carbon efficiency in electricity production and that the contributions from industry's voluntary efforts were relatively small (Domestic Institutions Subcommittee, Global Environment Committee, Central Environment Council, 2001). Indeed, energy intensity (consumption per unit output) by industry deteriorated considerably during the 1990s in Japan (Figure 10.1). The loose environmental policy network, thus, started to challenge the definition of the problem, and hence attempted to reframe the climate policy discourse.

At the time of writing there was no sign that Japanese GHG emissions would decline. The latest data showed that, in 2003, GHG emissions had increased by 8 per cent, and carbon dioxide

Figure 10.1 *Energy intensity in industry (energy consumption per unit output calculated on the basis of indices of industrial production which give a measure weighted for value added).*

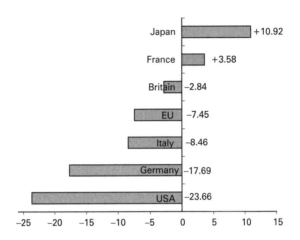

Source: Ayukawa and WWF (2001).

by 12.9 per cent, from 1990 levels,[2] meaning a 14 per cent reduction was necessary to attain the Kyoto target (*Yomiuri Shimbun,* 8 November 2004). The Kyoto Protocol was about to enter into effect (see Chapter 2) and the big gap between the EA and MITI (now the Ministry of the Environment and the Ministry of Economy, Trade and Industry, respectively) remained. The EA envisages a carbon tax and a voluntary domestic emissions trading scheme to be introduced between 2005 and 2007, while MITI has decided to make full use of the Kyoto mechanisms to attain the target (*Nihon Keizai Shimbun,* 4 November 2004). All peak industrial associations (Nippon Keidanren, the Japan Association of Corporate Executives and the Japan Chamber of Commerce and Industry) are opposed to the tax. Keidanren also opposes even voluntary domestic emissions trading, as it objects to government involvement in such a scheme and instead emphasises the contribution from its Voluntary Action Plan (*Asahi Shimbun,* 4 August 2004). I have looked at developments in Japanese global warming policy in general. Important developments have also taken place in integrating climate concerns into the energy and transport policy areas. Let me briefly explain the introduction of green automobile taxation, the 'greening' of energy-related taxation and spending, and the introduction of the renewable portfolio standards (RPS) system.

Green automobile taxation

The idea of green automobile taxation (GAT) aimed at reducing carbon dioxide emissions from the transport sector was first presented by the Ministry of Transport before COP3. One of the key advisers designing GAT was an environmental economist, Takamitsu Sawa, who was also advising the EA on tax reform to tackle the problem of air pollution as well as the carbon dioxide problem. With mediation by Sawa, the Ministry of Transport and the EA started to work together on GAT, focusing on the carbon dioxide problem with the aim of introducing it in the fiscal year 2000 (interview with Takamitsu Sawa, 2000). The finished proposal was to restructure car tax, light-car tax and vehicle weight tax. The tax burden would be reduced for new cars which met the latest fuel efficiency standards as set out in the Energy Conservation Law, while increasing the tax burden on those with cars whose fuel efficiency fell more than 20 per cent short of the old standards. Taxes on LEVs would be reduced but an increased burden would apply on diesel vehicles that did not meet the latest vehicle emissions standards. The Ministry of Transport and the EA estimated that GAT would reduce tax revenues in the first two years slightly, and would be neutral thereafter.

The proposal met with opposition on various fronts. MITI, which had just introduced a tax incentive measure to promote cars meeting new fuel efficiency standards, was strongly opposed to it, and argued the car manufacturers' concern that the measure would reduce the sales of the more profitable medium-sized and large cars when the industry was in recession. The MOF had reservations about the revenue losses in the first two years. The MOC was concerned about the impact of the GAT on the revenue generated by the vehicle weight tax, which was earmarked for road-building. The Ministry of Home Affairs was also reluctant to accept the proposal because the light-car tax was under its jurisdiction. While those car manufacturers that had put effort into increasing the fuel efficiency of their vehicles were in favour of such a tax measure (*Yomiuri Shimbun*, 22 May 1999), the Japan Automobile Manufacturers' Association was opposed to it, especially the increased tax burden on less fuel efficient cars. In doing so, the Association collectively protected the interests of the most vulnerable companies (interview with Takamitsu Sawa, 2000).

Although opposition was strong, the proposal for GAT gradually built up support from the media and the public, not least because of the Ministry of Transport's and the EA's efforts to raise awareness. In a bid to overcome opposition from the MOC and the construction *zoku* in the LDP, the vehicle weight tax was removed from the proposal and this enabled the construction, transport and environment committees within the ruling parties to agree to give it approval. Nevertheless, the proposal was then effectively shelved, as it was passed to the government's and the LDP's tax commissions for further consideration. The MOF and MITI remained opposed to it, while the removal of the vehicle weight tax from the proposal invited new criticism questioning its effectiveness.

Towards the next tax reform, the Ministry of Transport, the EA and, this time, MITI got together to draw up a new proposal on GAT. This version was approved by the MOF and ruling parties, and introduced in April 2002. The new GAT was considerably different from the previous design. First, it aimed both to improve fuel efficiency and to reduce emissions of nitrogen oxides and small particulate materials from vehicles. Second, the automobile tax alone was reformed. The automobile tax was a prefectural tax and some local authorities, including Tokyo, were already moving towards a greening of taxation. Taking into account the strong opposition from the MOC and in line with the previous year's agreement struck among ruling parties, vehicle weight tax was omitted from GAT. Third, and most important, it was designed to encourage people to replace old cars with new cars. The automobile tax would be reduced by up to 50 per cent, according to the car's environmental performance, for two years from the date of purchase. Cars' environmental performance was evaluated in relation to the new standards under the Automobile Nitrogen Oxides Law and the 'top-runner' standards under the Energy Conservation Law. At the same time, the tax burden would be raised by 10 per cent for cars more than thirteen years old, and for diesel cars eleven years old or more. Car manufacturers not surprisingly welcomed this measure, while environmental groups doubted its effectiveness as a green tax since environmentally harmful cars, if not old, were not subject to higher tax rates (*Kiko Network Newsletter* 79, 20 December 2000).

Combined with the preferential treatment of the vehicle-acquisition tax and a new subsidy programme for individuals purchasing LEVs like Toyota's hybrid car, Prius, GAT proved effective in terms of promoting the sales of efficient and less polluting cars. The low-emissions labelling scheme which the Ministry of Transport started in 2000 also raised environmental awareness among the public and helped promote these cars. According to the Ministry, in the first half of financial year 2002, 57 per cent of new cars sold had high environmental performance (*Kumanichi Shimbun*, 30 November 2002). The fuel efficiency of individual cars rapidly improved after Toyota marketed Prius just before COP3, aided by various financial measures, the labelling scheme, the 'top-runner' approach and Prime Minister Junichiro Koizumi's enthusiasm for LEVs. However, the general trend towards larger cars has not halted. GAT is, after all, not designed to reverse the 1989 tax reform which drastically cut taxes on larger cars.

Consensus policy styles allow veto players to distort policy rationality. Policy innovations are also bounded by existing policy frameworks, which are the past political outcome (Weir, 1992). Nevertheless, or therefore, GAT is strongly a positive-sum measure. It encapsulates the negative as well as positive effects of consensus corporatist policy-making, just as Keidanren's voluntary action was both pre-emptive (in the sense that it was designed to lessen the burden on industry by setting voluntary standards before government sought to do so) but also acted to some degree to make policy more concerted.

Greening energy-related taxation and spending
Following the heated debate over GAT at the end of 1999, the MOF's Tax Commission started to take the idea of environmental taxes seriously. In its report published in July 2000, it pointed out the need to examine the relationships between existing energy-related taxation and a carbon tax, and indicated its willingness to examine the matter further (Tax Commission, 2000). This laid the ground for political debate on the introduction of environmental taxes. The biggest opposition party, the Democratic Party,

included the introduction of an environmental tax as an electoral pledge for the election to the House of Representatives in 2000. MITI also quickly responded to the Tax Commission's move and in its report of July 2000 it argued for the need to review energy and road vehicle taxation and expenditure. In July 2001, MITI for the first time incorporated the idea of a carbon tax in its Long-Term Energy Outlook and proposed a greening of the Energy Special Accounts (ANRE, 2001).

The debate gathered the pace with the inauguration of the Koizumi Cabinet in April 2001. Koizumi launched a full-scale structural reform of taxation and expenditure. Earmarked revenues, especially those for road improvement, were targeted for reform. It was widely believed that these revenues, which generated vested interests, were a cause of policy sclerosis and the financial and economic problems Japan was facing. When the deliberations on these reforms started, there was an expectation that the introduction of environmental taxes would be discussed in detail. The idea was floated of reforming the earmarked taxes for road improvement to establish an environment tax (*Mainichi Shimbun*, 27 May 2001). It is worth noting that the MOF's Tax Commission, which contributed to the deliberations on the tax reform, was headed by Hiromitsu Ishi, who once chaired an EA group that had studied an environment tax. As the debate went on, environmental taxes proved not necessarily to be a priority. Nevertheless, they were still listed in the Basic Policies for Economic and Fiscal Policy Management and Structural Reform 2002 decided by the Cabinet in June 2002. Also, in May 2002, when an amendment to the Global Warming Law was passed, a supplementary resolution was attached that urged discussion of the introduction of environmental taxes within the framework of the broader tax reform. Meanwhile, environmental groups pushed the debate further. The Japan Center for a Sustainable Environment and Society, together with the Kiko Network, the WWF, tax accountants and researchers, created a study group to conduct research, stimulate national debates and propose a carbon tax.

In June 2002 the government announced its decision to reform the Energy Special Accounts, which were principally under MITI's jurisdiction. Specifically, the proposal was to reclassify some of the revenues for energy policy from earmarked energy taxes as general account revenues. Detailed revisions were to be discussed at the Prime Minister's Council on Economic and Fiscal Policy, which had considered the tax and expenditure reforms (*Yomiuri Shimbun*, 10 June 2002). In August MITI set up a group to review its Energy Special Accounts, and within a few weeks the Minister, Takeo Hiranuma, announced, at the Council on Economic and Fiscal Policy, an energy tax reform which included a provision to tax oil, natural gas and coal partly according to their carbon dioxide emissions (*Asahi Shimbun*, 21 October 2002).

The new taxation has been applied since October 2003. The main points of the reform include:

- A tax on coal is being phased in between 2003 and 2007.
- For oil, LNG and LPG, in addition to an import tax, a new tax based on carbon dioxide emissions has been introduced. Taxes on LNG and LPG are set to increase between 2003 and 2007, but the tax on oil has been frozen.
- Revenues from taxes on carbon dioxide emissions are earmarked for energy conservation and 'new energy' sources to tackle global warming. Some of these revenues are at the EA's disposal.
- Taxes on electricity have been reduced so as to keep the reform revenue neutral.
- Coal used to manufacture iron and steel, coke and cement, and for electricity generation in Okinawa prefecture is exempted (unitl April 2007).
- The funding for energy/environment policy in the two Energy Special Accounts has been merged to create a single Energy/Environment Special Account.

MITI's proposal announced by Hiranuma raised controversy within the EA. The proposal was not only for a greening of the funding from the Energy Special Accounts, the necessity of which both the EA and MITI recognised, but also effectively for a carbon tax – the EA's pet policy. With the new Climate Policy Programme, which stipulated a step-by-step approach (see above), the EA envisaged the introduction of a carbon tax after 2005, and its Central Environment Council had just released an interim report to that effect (Special Committee for Global Warming Prevention Tax, Central Environment Council, 2002). In November, MITI and the EA unusually released a joint statement about MITI's tax reform proposal to clarify three points: that the new taxes based on carbon dioxide emissions were not an environmental tax, or specifically a global warming prevention tax; that the introduction of a global warming prevention tax would be examined, but in accordance with the step-by-step approach; and that the future review of energy tax rates would be discussed between MITI and the EA (Ministry of Economy, Trade and Industry and Ministry of the Environment, 2002). A liaison conference for energy and environmental policy was also established between MITI and the EA.

Environmental groups welcomed the move, but stressed the need to introduce a separate carbon tax, and criticised the low tax rate on carbon dioxide emissions. Compared with the EA's latest proposal for a tax of ¥2,400 per tonne of carbon – which was still criticised as too low – MITI's tax was ¥1,070 (*Yomiuri Shimbun*, 5 November 2004; *Nihon Keizai Shimbun*, 23 April 2003; *Kyoto Shimbun*, 14 November 2002). Industry (Keidanren and the iron and steel, electricity, gas, chemical, and pulp and paper industries) opposed the tax. Nevertheless, the bill to reform the Energy Special Accounts was passed by the Diet in April 2003 – less than a year since it was first proposed. Exemptions, the freeze of taxes on oil, a generous cut in the tax on electricity and relatively low tax rates perhaps eased opposition from those industries which had strongly resisted the EA's proposal for a carbon tax. At the same time, we should not overlook the relationships between MITI and industry. The EA failed to have regular meetings with Keidanren on global warming policy, specifically over the introduction of a carbon tax (*Denki Shimbun*, 2003). That MITI had good negotiation channels with industry appears an important factor. Moreover, MITI and major industrial representatives remained united against the EA's proposed carbon tax. Indeed, industry now found another reason to oppose the EA's carbon tax. The power industry, for example, claimed that the tax would result in double taxation (*Nikkei Sangyo Shimbun*, 29 August 2003). According to a pro-industry newspaper, it was widely thought among key ministries and industrial sectors that the EA's argument for the need to introduce a carbon tax after 2005 lost its validity (*Denki Shimbun*, 20 February 2003).

What stands out in policy developments on the greening of energy-related taxation and expenditures is how rational MITI's move was. By reforming its Energy Special Accounts, MITI doubly co-opted potential sources of interference in its policy area: it

defended its interests in earmarked energy revenues and its jurisdiction over energy-related global warming policy. Nevertheless, it
was still an important policy development.

This case shows that important policy changes can take place
quickly in Japan. Whatever the motives, the leadership role taken
by MITI was the key factor, but close relationships between MITI
and key industrial actors also appear important. These factors,
however, have to be understood in terms of specific political conjunctures which gave rise to particular perceptions of what was
rational, and which opened up a new policy opportunity.

Renewable portfolio standards
After 1999 notable political developments took place in the promotion of 'new energy'. In November 1999 politicians from
all parties except the Communist Party formed the National
Parliamentarians' Association for Promoting Renewable Energy,
which was headed by a former Director-General of the EA, Kazuo
Aichi. In 2000 the Association had more than 250 members and
from that August the former Prime Minister Ryutaro Hashimoto
chaired it. In parallel, in May 1999, the Renewable Energy
Promoting People's Forum set up the Green Energy law Network
(GEN), which brought together about twenty NGOs with the aim
of introducing a law which would require power companies to
purchase a fixed amount of power generated from green energy
at fixed prices; a similar law in Germany had proved very effective. In 2000 the GEN joined the Parliamentarians' Association
and together they drew up a bill to promote green energy. Owing
to strong opposition from the electricity industry and trade union,
the bill did not stipulate a purchase obligation; instead, it would
require power companies to submit their purchase plans to MITI,
which could demand changes to the plans if deemed necessary.

This political development stimulated MITI, the power
industry and the LDP's nuclear energy *zoku*. MITI established
the New Energy Subcommittee of the Advisory Committee
for Energy, which included in its members Tetsunari Iida, the
representative of the GEN, and in May 2000 it started to examine
a new measure to facilitate the expansion of the markets for
new energy. As an alternative to the proposal from the GEN
and the Parliamentarians' Association, MITI floated the idea of
an RPS system – a measure similar to the British market-based

system (see below). Under this system, power companies would be required to include a minimum proportion of renewable energy in their portfolio of electricity resources. The required percentage of renewable energy would grow over time. Power companies have great flexibility in meeting this standard. They can develop 'new energy' power sources by themselves, purchase such energy or buy tradable certificates which are issued by government to green power generators in accordance with the amount of green electricity they generate.

The power companies resisted government intervention and a new burden on them, especially when liberalisation of the electricity markets was taking place, and in an effort to forestall any new mandatory measures they established the Green Power Fund and the Green Power Certificate System in October 2000. These were financial programmes to assist green energy developers through voluntary contributions from both individual customers and companies. The industry insisted that new energy should be promoted on a voluntary basis through these programmes.

Meanwhile, the LDP's nuclear energy *zoku* was concerned that green energy was gaining support from the public and that this might be translated into anti-nuclear sentiment. Several serious nuclear accidents in Japan from 1995 had brought about a crisis in policy on nuclear energy. The nuclear energy *zoku* appears to have been concerned also about the provision in the Parliamentarians' Association's bill to support green energy financially by using some of the earmarked taxes on electricity, about 80 per cent of the revenues from which were used to promote the development of nuclear power stations (*GEN Newsletter*, 17 March 2000). The concern about the effect of new energy promotion on nuclear energy promotion was shared by the power companies. During the deliberations of the New Energy Subcommittee, some members from the electricity companies stressed the need to state clearly that the expansion of new energy did not require a move away from nuclear energy (New Energy Subcommittee, Advisory Committee for Energy, 2000).

Opposed by the power industry and trade union and the LDP's nuclear energy *zoku*, the Association's bill had to be shelved. Meanwhile, MITI's RPS scheme was gaining support. Economists on the New Energy Subcommittee and its subcommittee designing a new measure to expand the markets for new energy were strongly in favour of RPS. The ruling parties' green energy project team and the LDP's energy committee expressed their support for it. In such a political climate, some politicians in the Parliamentarians' Association drew up an amendment to the draft bill basically applying the RPS approach. Power companies were resistant to the RPS system, and still emphasised their voluntary programmes; however, their concern quickly shifted to the design of an RPS system once MITI started to discuss its details.

As the specific design of the RPS bill emerged, the GEN, together with other environmental groups, renewable energy producers and environmental think-tanks, set up the Citizens' Green Energy Committee in order to scrutinise the debate. Environmental groups, including the GEN, the Kiko Network, Greenpeace and FOE, together with green energy producers, also held a joint press conference, at which they called for revisions to the emerging bill. Their activities certainly raised the public's interest in the issue (*GEN Newsletter*, 20 March 2002). Meanwhile, the Parliamentarians' Association prepared two draft bills and, instead of submitting them to the Diet, handed them to MITI, which was hammering out the RPS bill.

The Diet passed the RPS Bill in May 2002 and the law came into effect in April 2003. During the Diet debate, the opposition parties submitted an alternative bill and raised again the controversial issue previously discussed by the New Energy Subcommittee of whether energy from waste should be included in the definition of new energy. In the end only biomass was to be counted. The details of the law, including the overall target and annual targets for power companies, were revealed in MITI's ministerial ordinance issued in February 2003. The overall target was to source 1.35 per cent of the total electricity supply from new energy by 2010, or 12.2 TWh, compared with about 3 TWh in 2002.[3] Environmental groups criticised this target as unambitious while the power industry complained that it was too heavy a burden. Overall, however, the industry appears to have been effective in its efforts to reduce the burden: through various provisions the power companies were allowed virtually to maintain their existing proportion of new energy sources for the first five years.

This case shows that the politics of energy and the environment is increasingly pluralist in the sense that various actors come into play in the policy-making process. As seen in the alliance between environmental groups, green energy generators and politicians, political processes were more dynamic than they typically had been. The exclusive corporatist policy-making arena began to open up for environmental groups. Nevertheless, it is too rash to conclude that Japanese energy/environment politics entered a new phase. Iida from the GEN, for example, was excluded from the subcommittee that considered the detailed design of the RPS law. In this sense, the incorporation of Iida in the New Energy Subcommittee was more a political gesture. Indeed, the GEN could not really break the existing exclusive corporatist nature of energy policy-making and had only a limited influence on policy contents. Nevertheless, the political and civil activities of the GEN attracted the interest of the mass media and raised public awareness of the issue. Prominent environmental groups worked together and their activities were well coordinated. The GEN certainly laid the foundations for future developments in environmental campaigning and energy/environment politics in Japan.

Summary

It is safe to argue that MITI and industry (especially Keidanren) had the leading role in shaping global warming policy in Japan after COP3. The political landscape was characterised by rivalry between a loose coalition centring the EA and the corporatist close network centred on MITI, and policy developments took place in a way that allowed MITI and industry to forestall the EA. In one sense the EA's challenges to the MITI–industry network generated or brought to the fore the common interests between MITI and industry in taking proactive action to maintain the status quo in their relations and control over their affairs. Of course, this picture of MITI–industry-led climate politics is a rough one. MITI and industry do not necessarily share interests and industry is not a monolith. Also, as seen especially in the politics of GAT and the RPS law, politics is more dynamic than before. Nevertheless, this picture largely underlies Japanese global warming politics and policy before and after the Kyoto conference.

When Keidanren merged with Nikkeiren (Japan Federation of Employers' Associations) to create Nippon Keidanren (Japanese Business Federation) in May 2002, there was an expectation that this might bring about changes in the attitudes of the most important and influential representative of industry. Takashi Imai from

Keidanren and Nippon Steel was replaced in his post of 'the prime minister in the business world' by Hiroshi Okuda from Nikkeiren, which had been relatively positive about global warming policy. Moreover, Okuda, who had once been engaged in the development of hybrid cars at Toyota, was seen as relatively receptive to the idea of an environmental tax (*Mainichi Shimbun*, 11 June 2002). In June 2002 Okuda showed a positive attitude towards an examination of an environment tax (Nippon Keidanren, 2002), which Keidanren – dominated by the power, iron and steel and other energy-intensive industries – had resisted (*Mainichi Shimbun*, 11 June 2002). This inflated an expectation that industry might ease its opposition to a carbon tax. Nevertheless, Okuda, as the head of Nippon Keidanren, in the end announced its firm opposition to an environment tax. This indicates how the interests of energy-intensive industries have prevailed and have influenced Japanese global warming policy.

Policy developments in Britain

In May 1997 the Labour Party came to power promising to be 'the first truly green Government ever' and to put 'concern for the environment at the heart of policy making' with its manifesto commitment to cut carbon dioxide emissions alone by 20 per cent by 2010 from 1990 levels (New Labour, 1997: 39). Labour also strengthened the British 'burden-sharing' target for the EU's overall target for GHG reductions under the Kyoto Protocol. Before the Kyoto conference, the Conservative government had agreed to the British target of a 10 per cent reduction in GHG emissions, to contribute towards the EU's 15 per cent reduction target proposed for the COP3 negotiations. Labour, however, accepted a 12.5 per cent reduction to contribute to the EU's target of only 8 per cent under the Protocol. In terms of carbon dioxide emissions, this amounted to a reduction of about 8.5 per cent. There was speculation that Labour might drop its manifesto commitment of a 20 per cent reduction in carbon dioxide emissions when it came to power; however, it reaffirmed that it would 'move towards' a 20 per cent reduction as a domestic target in *Climate Change: The UK Programme*, published in November 2000 (DETR, 2000).

In 2003, in its second term of office, Labour published an energy white paper that was based on the reports issued by the RCPE in 2000 (RCEP, 2000) and by the Cabinet Office Performance and Innovation Unit in 2002 and a major consultation conducted by the government itself in 2003. The white paper, entitled *Our Energy Future – Creating a Low Carbon Economy*, was a watershed in energy/environmental policy in Britain. Priority was given to environmental objectives and the role of renewables and energy efficiency as a strategy was endorsed. Especially notable was a commitment to a 60 per cent reduction in carbon emissions by 2050, with significant progress to be made by 2020 (DTI, 2003). The target was in a direct response to the RCEP report on energy and climate change. The House of Commons Environment Audit Committee (2003: para. 11), although it was concerned about the lack of specific measures, welcomed this strategic vision and stated that Britain had 'led the way internationally'.

Labour's climate change policy was more ambitious than the Conservatives' programme, which was characterised by reliance on voluntary action. The new climate change programme emphasised 'gains not pain' from the climate policy by positively linking its measures with social and economic objectives; the government declared it was seeking a 'first mover' advantage for British industry and the economy. The new programme outlined measures that would reduce carbon dioxide emissions by 19 per cent and total GHG emissions by 23 per cent. The remaining 1 per cent towards the domestic target of a 20 per cent reduction in carbon dioxide emissions would be met with measures such as further action by the devolved administrations, local authorities and a public awareness campaign. The key features of the programme were: the climate change levy (CCL), climate change levy agreements (CCLAs); a new Carbon Trust to facilitate the take-up of cost-effective low-carbon technologies; a domestic emissions trading scheme; the introduction of a Renewables Obligation; and a Ten-Year Plan for Transport. At the time of writing, the climate change programme is under review. The government planned to publish a revised programme by the end of 2005.

Prime Minister Tony Blair has kept up momentum on climate policy and sent a signal that the policy will stay and be strengthened – not eased – in several speeches devoted to the environment and global warming. Britain proved to be one of the few countries to have met its GHG reduction target for 2000 – stabilisation of carbon dioxide emissions at 1990 levels. Indeed, it went much further. Between 1990 and 2000 carbon dioxide emissions declined by 8 per cent. With sizeable reductions also in its emissions of other GHGs, Britain made a 12.8 per cent reduction in its overall emissions over the same period, meaning that it has already achieved the binding target under the Kyoto Protocol. In policy terms also, Britain became the world leader by introducing the first economy-wide domestic emissions trading scheme, in 2002.

Despite Britain's apparent success in reducing GHG emissions, the ambitious long-term target (from the energy white paper) of a 60 per cent reduction in carbon emissions and the evident political commitment, carbon dioxide emissions in fact had risen slightly between 1997 and 2004. The House of Commons Environment Audit Committee (2004: para. 21) expressed its concern that the government's climate policy was 'seriously off course'. The reduction in carbon dioxide emissions was largely the legacy of electricity privatisation under the Conservatives, which gave rise to the 'dash for gas' (see Chapter 7).

This contrast between success and innovation, on the one hand, and stagnation, on the other, manifests itself in the development of policies and measures. This is looked at below in the context of the CCL package, emissions trading and transport policy.

Climate change levy package
The CCL was introduced in April 2001, after long consultation, as a central plank of the Labour government's climate change programme as well as of the first major project based on the Statement of Intent on Environmental Taxation issued by the Treasury in July 1997, which professed the government's new policy direction to shift the burden of taxation from 'goods' (e.g. employment) to 'bads' (e.g. environmental pollution). The CCL is a downstream tax on business energy use. Natural gas, coal and electricity used by business, agriculture and the public sector are all subject to the tax, but electricity generators are exempt. This design reflected the government's desire to protect both those in a state of 'fuel poverty' (see Chapter 7) and the coal industry. The CCL was a policy package linked with several other policy initiatives (see Box 10.1).

The introduction of the CCL presents a major break from the Conservatives' policy, which strongly resisted a carbon or energy tax on industry out of concern for its effect on economic competitiveness and under strong business lobbying. Why and how could the policy change? Several factors were instrumental in enabling Britain to introduce the CCL.

First, Labour needed a policy to deliver the British GHG reduction target under the EU burden-sharing scheme and its own target set in its 1997 election manifesto. While committing Britain to an ambitious carbon dioxide reduction target, Labour nevertheless cut the 8 per cent VAT on domestic energy to 5 per cent – a minimum allowed by EU law, as promised in its election manifesto. The government also placed a moratorium on further construction of gas-fired power plants to rescue the coal industry (Advisory Committee on Business and the Environment, 1998: 5).

Second, and more important, key political leaders were in favour of the use of economic instruments. The Chancellor,

---

Box 10.1 *Key features of the climate change levy package*

Climate change levy
• The levy would be a tax on business use of electricity, gas and coal. A 10–15 per cent rise in energy prices was expected.
• Renewable energy and good-quality CHP would be exempted from the levy.
• Domestic energy use would also be exempted.

Earmarked use of revenue from the levy
• A reduction in National Insurance contributions by 0.3 per cent.
• Capital allowances: 100 per cent first-year capital allowances for energy-saving investments by the private sector.
• The Carbon Trust (an independent non-profit company to promote low-carbon technologies and innovation).

Climate change levy agreements
• There would be an 80 per cent rebate for members of industrial associations in energy-intensive industries that voluntarily entered into negotiated agreements with the Department of the Environment, Transport and the Regions on energy efficiency or carbon reductions. Eligibility to negotiate agreements was limited to those covered under the EU IPPC (Integrated Pollution Prevention and Control) Directive.

Gordon Brown, the Deputy Prime Minister and the political head of the newly established Department of the Environment, Transport and Regions (DETR), John Prescott, and the Environment Minister, Michael Meacher, were all enthusiastically in favour of environmental taxation. An important point here is that they supported environmental taxes with the hypothecation provision. The hypothecation provision was one of the key factors needed to make environmental taxes more politically acceptable, but the Treasury had been persistently opposed to the idea of any sort of hypothecation for fear of losing its control over budgeting. The CCL overcame this constraint. It was proposed that the revenue raised would be recycled back to business, largely through a reduction in employers' National Insurance contributions, but also by earmarking part of the revenue for financial incentives for investment in energy efficiency.

In opposition, Labour was initially against the use of economic instruments, as it was concerned about their effects on the poor. It is widely believed that the party was strongly influenced by reports advocating the use of economic instruments by the Institute for Public Policy Research (IPPR). A report published in 1996, in particular, refuted the argument that environmental taxes hit the poor most and argued that environmental taxation provides a 'double dividend', that is, a positive effect on both the environment and employment (Tindale, 1996). Inherent in this argument is the interventionist role of the government in giving rise to positive relationships between environmental policy and the economy. The argument that well designed environmental measures can stimulate the economy appears more acceptable to Labour than to Conservatives committed to a neo-liberal, non-interventionist political ideology.

Another important factor that enabled Labour to introduce the CCL was authoritative acceptance of the need for it. In 1998, in response to the request by the Prime Minister's Office, the Advisory Committee on Business and the Environment issued a report on the role of business in the government's climate policy. The report favoured negotiated agreements and was positive about an emissions trading scheme. At the same time, it reluctantly admitted the necessity of a tax, with some conditions, including that the revenue raised would be fully recycled by helping industry to reduce carbon dioxide emissions and/or by being revenue neutral, and that business competitiveness would not be damaged (Advisory Committee on Business and the Environment, 1998: 10–11). In his 1998 budget, the Chancellor asked Lord Marshall to lead a government taskforce on an energy/carbon tax on industry and consider whether economic instruments should be used to improve energy efficiency in business, and, if so, how. Lord Marshall was at the time the president of the CBI. The design of the tax was, therefore, to be proposed by a prominent business leader.

In nutshell, the 'policy window' opened (Kingdon, 1984). There were political willingness and leadership. There was a policy need. There was also authoritative acceptance of its necessity. Lord Marshall concluded his report in November 1998 and in the 1999 budget, which represented a watershed in British green tax reform, the Chancellor announced the introduction of the CCL, to be effective from April 2001.

The CCL, however, proved very unpopular among businesses. The government stuck to its decision to introduce the CCL, but had to make various concessions in the face of strong business lobbying. The tax rate was cut by almost 30 per cent from the original proposal. The 50 per cent tax rebate, which the original proposal made available to 'energy-intensive' sectors in exchange for achieving negotiated voluntary targets for energy efficiency or carbon reduction (the CCLAs), was extended to 80 per cent. The definition of 'energy-intensive' sectors also made the proposed CCL susceptive to business criticism. The DETR, which was responsible for the CCLAs, categorised installations regulated under the EU's Integrated Pollution Prevention and Control Directive as 'energy intensive'. As a result, many non-energy-intensive industries were given the discount while some industries were excluded that had a higher energy intensity than many with access to the CCLAs. When the CCL was introduced as many as forty-four industrial sectors were included in CCLAs (*ENDS Report* 319, 2001: 16).

Even after the CCL was introduced, business, in particular the CBI and the Engineering Employers' Federation (EEF), continued lobbying the government to reform the CCL. Despite being designed to be revenue neutral, largely through the reduction in National Insurance contributions, the scheme hit the manufacturing sector disproportionately hard. In November 2002 the CBI and the EEF launched a renewed campaign and urged the Treasury to extend CCLAs linked to an 80 per cent levy rebate to a wide cross-section of businesses. The continued industrial lobbying brought fruits in 2003. In the 2003 pre-budget statement, the Treasury finally promised to widen eligibility for CCLAs with a levy rebate, giving up an earlier determination not to give in over the eligibility criteria (*ENDS Report* 347, 2003: 26). An extra £5 million was also made available for enhanced capital allowances for energy-saving technologies (*ENDS Report* 339, 2003: 4). At the same time, the levy rate has never been raised since its introduction.

It is notable that when the Conservatives pledged the abolition of the CCL during the 2001 general election campaign, this did not attract the full support of business. The EEF, in particular, saw the need to reduce GHGs and was more concerned about the design of the levy package (Major Energy Users' Council, 2001). At the same time, the EEF's campaign to widen the eligibility criteria for CCLAs indicates the level of stringency of the targets under the agreements. According to the ETSU, the CCLAs represent about 60 per cent of the 'cost-effective' potential for energy efficiency beyond 'business as usual' (ETSU, 2001).

The CCL was also criticised by environmentalists. The RCEP, for example, urged the government to replace the CCL with a carbon tax related directly to carbon dioxide emissions. The gas and electricity regulator, the Sustainable Development Commission and the IPPR were among those who shared that view. The problem here was that a tax on carbon would affect household energy bills. After the Conservatives' failure to introduce full-rate VAT on domestic energy, politicians were more sensitive about the problem of fuel poverty. The CCL was an innovation within the 'boundaries' created in the past (Weir, 1992).

What the politics of the CCL suggests is that the most affected groups are effective in emasculating government policy. By arguing this, I do not mean that the CCL package did little to cut carbon dioxide emissions. The levy enabled the government to negotiate a number of agreements which the Conservative administration had largely failed to introduce in the 1990s; according to the CBI and the EEF, these agreements were effective in encouraging the efficient use of energy. The few veto points available to British institutions indicate strong institutional capabilities to introduce innovation. Notwithstanding, economic interests were effective in defending their interests, thus undermining policy rationality.

Britain as a pioneer: emissions trading in Britain and the EU
In January 2002 the world's first economy-wide domestic emissions trading scheme opened in Britain. The government's expectations were high for the potential benefits arising from the establishment of the City of London as a centre of a new international market, as the Kyoto Protocol had included a provision for international emissions trading as a means to reduce GHG emissions. It was, however, business that led the debate and policy formulation in this initiative. Following the recommendation for a pilot emissions trading scheme in the 1998 Marshall report, in June 1999 the CBI and the Advisory Committee on Business and the Environment created the Emissions Trading Group (ETG), which brought together around 100 major companies, trade associations and other organisations, to develop a scheme as an *alternative* to taxation. An emissions trading scheme had been preferred by business as the most cost-effective means to reduce GHG emissions, and energy-intensive industries and power producers were calling for a trading scheme instead of a carbon/energy tax. The government took part in the work of the ETG essentially as an adviser, and in March 2000 the ETG outlined its proposal for a scheme. The government then published a consultation paper in November 2000 and the rules were finalised in August 2001.

The pilot scheme was a voluntary one, which both industry and the government (the DTI and the DETR) preferred. Firms could join the trading scheme by bidding on absolute emission reduction targets. To entice firms to take the risk of volunteering absolute caps, or limits, on their emissions, a total of £215 million was granted as a financial incentive. The government initially suggested voluntary entrants offer at least a 1 per cent per year reduction in emissions, but this was dropped in the face of the ETG's opposition. To ensure liquidity for the scheme, it was linked with the CCL package: firms under CCLAs were allowed to buy permits to meet their targets or sell over-achievement into the scheme. Other firms were also given access to the trading scheme as sellers of credits generated from emission reduction projects within and outside Britain. The government proposed to restrict access to these project credits to firms with absolute targets, thus excluding those in CCLAs with output-based targets. However, the ETG won concessions from the government so that all participants could buy these credits to meet their targets.

The trading scheme, by linking with CCLAs, became very complicated and also affected the provisions of CCLAs. Crucially, it affected the effectiveness of CCLAs. Most firms under CCLAs had targets relative to their output and these targets were based on an evaluation of the potential for cost–benefit measures to be implemented; however, firms were now effectively provided with a way to eschew investing in these measures (*ENDS Report* 319, 2001: 17).

In March 2002 thirty-four organisations took on absolute emission reduction targets at an auction. Emissions reduction from the auction was 60 per cent more than officially expected. According to the Department for Environment, Food and Rural Affairs (DEFRA), the targets represented an 11 per cent reduction in emissions, on average, from the baseline, which was based on the average annual emissions from 1998 to 2000 (*ENDS Report* 326, 2002: 25).

The British emissions trading scheme, however, soon faced problems. First, according to the *ENDS Report*, several big participants, especially from the chemical industry, were found to be likely to gain from emission reductions which had already been achieved under regulatory requirements, and with generous windfall payments from the government's £215 million incentive fund (*ENDS Report* 326, 2002: 25–9; 327, 2002: 3–5). Second, the emerging EU emissions trading scheme, due to start in 2005, would cut across the British scheme in several respects. Among the most important was that the proposed EU scheme included electricity generators in its list of mandatory participants. In order to protect the fuel poor and the coal industry, the British trading scheme, and the CCL and CCLAs linked to it, basically excluded electricity generators, by passing emissions from electricity generation to business consumers of electricity. The EU scheme, to the contrary, attributed emissions from electricity generation to the generator. This had serious implications for the future of the CCL and CCLAs, let alone domestic emissions trading.

As a pioneer of economy-wide GHG emissions trading, however, the government was enthusiastic in taking a leadership role in the EU emissions trading scheme and the ETG was closely involved in the British engagement in the EU debate and scheme.

One of the key issues surrounding emissions trading was the allocation of emissions allowances. The EU scheme basically left it to member states to decide how many allowances should be issued for industry covered by the scheme, although the Commission retained a right of veto over national allocation plans, to ensure effectiveness of the scheme. Governments were, specifically, asked first to cap the overall carbon dioxide emissions from industry and then to parcel out the total allowances between sectors and individual installations. Britain had already achieved its 'burden-sharing' target under the Kyoto Protocol but it was far from meeting its own domestic target of a 20 per cent reduction in carbon dioxide emissions. The ETG, representing industry, strongly lobbied the government to use the Kyoto target as a basis for allowance allocation, in order not to disadvantage British industry. Environmental groups, on the other hand, pushed the 20 per cent domestic target, especially given the ambitious target of a 60 per cent reduction in carbon emissions as stated in the energy white paper published in 2003.

In January 2004 the government published its draft national allocation plan, the first among member states to do so (DEFRA, 2004b). The draft plan represented a victory for environmental groups. Overall allocation of allowances for the scheme's first phase (2005–7) was intended to reduce carbon dioxide emissions slightly further than what would have been achieved with existing policies and measures, still approaching the 20 per cent reduction target. Specifically, the target for the first phase was aimed at contributing an overall 16.3 per cent reduction in carbon dioxide emissions, against the business-as-usual forecast of a 15.3 per cent reduction between 1990 and 2010. The manufacturing sector's allowances would be based on business-as-usual projection; however, the extra reduction was to come from the power generation sector, as it was less exposed to international competition. These additional savings equated to 5.5 million tonnes of carbon dioxide in total from 2005 to 2007. By 2010 the power generation sector was expected to make a reduction of 5.5 million tonnes of carbon dioxide a year, consistent with the contribution of emissions trading to carbon dioxide reduction stipulated under the British climate change programme.[4] In the second phase (2008–12), the allocation would be tightened so as to ensure that industries covered by the scheme would make due contributions to achieving the 20 per cent reduction target. The EU Directive on emissions trading required that governments should not allocate more allowances than were likely to be needed (EU Directive 2003/87/EC, Annex III). Basing the allocation on Britain's burden-sharing target would go against the Directive's intention, and such an arrangement might therefore invite the Commission's objection and rejection of the plan. Britain, which was keen on setting the pace of the EU scheme, chose to be ambitious.

This draft allocation plan provoked fierce industrial opposition, especially because the British plan proved to be among the most stringent in terms of its ambition to go beyond the Kyoto target, while most other major member states were hesitant to use the scheme as a tool to ensure achieving their burden-sharing targets.[5] The government stuck to its plan, nonetheless, and submitted it to the Commission at the end of April 2004 and published it in May (DEFRA, 2004c).[6] Environmental groups and DEFRA pressed the European Commission to take a tough line on other member states' national allocation plans, and notably the CBI joined this move.[7] The Prime Minister himself expressed concern to the Commission President (*ENDS Report* 354, 2004: 56–7). Meanwhile, industry also continued to lobby hard; the DTI was asked for revisions to the forecast carbon dioxide emissions, and hence increases in emission allowances.

In October 2004 DEFRA announced the government's intention to submit a new plan following revisions to the emission forecasts. The new forecast issued by the DTI estimated 7.6 per cent higher emissions of carbon dioxide and overall emission allowances in the first phase were increased by 19.8 million tonnes of carbon dioxide to 756.1 million tonnes (DEFRA news release, 27 October 2004). DEFRA explained that the new allocation was some 5.2 per cent below the business-as-usual projections, whereas the original projections were 0.7 per cent below. The additional savings of 5.5 million tonnes of carbon dioxide from the EU scheme for the first phase were retained, but the savings were to come from a much more generous baseline. The CBI was not entirely satisfied with this concession, but welcomed the government's move to soften the allocation plan.

There was apparently a fierce fight between DEFRA, which defended the existing plan, and the DTI, which represented industry; in the end, the Prime Minister sided with the DTI and industry (FOE, 2004; Press Association, 27 October 2004). Having given a major speech on climate change just a month previously, this greatly undermined the Prime Minister's environmental credibility. It was also a humiliation for the British government, which had presented itself as leading the way in the EU emissions trading scheme, but had instead to ask the Commission for amendments to the plan at the last minute before the introduction of the scheme in January 2005. Nevertheless, with the next general election approaching, backing those with immediate, concentrated interests was a politically rational choice, especially over an issue which creates no immediate winners in economic terms and which generally fails to stimulate public interest.

These developments on emissions trading policy show the growing importance of the role of the EU in global warming policy. As the pioneer of GHG emissions trading, Britain wanted to lead the policy debate on EU emissions trading and it was of course in the British interest to do so. However, in the end, not only was British influence on the development of the EU scheme limited, but also the British policy was more influenced by developments at the European level than the other way round. This has certainly created considerable administrative costs, as domestic policy has had to adjust to EU policy, and, as seen in the policy retreat on the emissions allocation, political costs since the government's ambition was reduced by the lukewarm attitudes on the part of other European governments. Nevertheless, the EU emissions trading scheme also provided an opportunity. It was the first economic instrument introduced to tackle carbon dioxide emissions to have affected households since Labour came to power.[8] The government had been repeatedly urged to take action on this front, but shied away from this politically risky issue. The EU scheme broke the policy sequence since the Conservatives' failure to introduce the full rate of VAT on domestic energy.

Transport and the environment: fuel protests in 2000
One of the areas where the Chancellor, Gordon Brown, targeted in his green tax reform was road transport. In his first budget statement, in 1997, Brown announced his intention to raise the fuel duty escalator from an annual increase of at least 5 per cent (as had been implemented by the Conservatives) to 6 per cent in real terms. The main justification he gave was to the need to reduce

carbon dioxide emissions. In the budget in 1999, which the House of Commons Environment Audit Committee (1999b: para. 9) praised as 'greenest ever', the Chancellor raised duty on standard diesel by 11.6 per cent. This, however, was followed by a series of policy retreats.

In the context of rising oil prices, the increase provoked the haulage industry. The high fuel duties were also attacked by motoring organisations and the Conservative Party backed these moves and criticised Labour as 'anti-motorists'. In face of a fierce campaign against high fuel duties, the Chancellor was forced to make concessions in his 1999 pre-budget statement. He announced the scrapping of the 6 per cent fuel duty escalator and said that future increases in fuel duty on would be decided budget by budget (rather than at a fixed annual rate above inflation, as previously). In his budget in 2000, fuel duties were simply kept in line with inflation. A number of measures were also introduced to 'boost competitiveness' of hauliers, effectively encouraging road freight over rail. The Chancellor did continue his green tax reform on road transport by graduating vehicle excise duties for new cars and company car taxes according to their rate of carbon dioxide emissions, but the net effect was that the tax cost of car ownership was cut by almost £250 million in total (joint Treasury/DETR press release, 21 March 2000).

A further blow came with the 'fuel crisis' in the autumn of 2000. Following direct action against high fuel prices in France, which drew some concessions from the French government, hauliers and farmers in Britain launched a similar protest, demanding cuts in fuel duties in the pre-budget statement of 2000. Blockades of oil refineries and depots by protesters soon gave rise to a near crisis in fuel distribution and supply, causing massive socio-economic disruption. The protest received general support from the public as well as the leader of the Conservative Party, who criticised high fuel duties as a 'stealth tax' while promising a cut of 3p per litre on fuel duty as an electoral pledge for the forthcoming 2001 general election.[9]

During the protests, Labour tried to defend its fuel duty policy; however, the argument was based on the merits of raising revenue for, say, schools and hospitals, rather than environmental objectives. It was only John Prescott among Cabinet ministers who spoke of the link between environmental objectives and fuel duty policy (*ENDS Report* 308, 2000: 20). Environmental groups, including FOE, WWF-UK, Greenpeace and the RSPB, eventually stood up to claim the environmental case for high fuel duties and held a joint press conference.

In the end, the environmental groups lost the battle. In his pre-budget statement the Chancellor introduced a number of measures to reduce the overall tax burden on road hauliers and motorists, at a cost of £2 billion (Treasury press release, 8 November 2000). Nevertheless, the Chancellor, again, did not forget to paint his budget green (see Box 10.2).

This was a serious climb-down from the high-tone statement on environmental taxation presented by the Chancellor in 1997, to the effect that polluters should face the full costs of their actions as long as competitiveness was not damaged. The Freight Transport Association and the Road Haulage Association claimed that they were losing out to foreign competition because of the high fuel duties and vehicle excise duty, but the government maintained that the total tax burden – including corporation tax, social costs and tolls on continental roads – was not particularly heavy on British hauliers. Moreover, the House of Commons Environment, Transport and Regional Affairs Committee (2000b) indicated that vehicle excise duty and fuel duty should be increased so as to reflect environmental and social costs fully.

The government's retreat was not only on the tax issue. The much promised 'integrated transport policy' and the Ten-Year Transport Plan, a long-term investment strategy to implement it, were also affected by events. The government's Commission for Integrated Transport prompted the government not to shy away for fear of being seen as 'anti-car', and argued that doing nothing would be the true anti-motorist policy. However, having experienced the severe backlash from motorists and hauliers during the fuel duty protests, the government lost political will and courage, as well as vision (Commission for Integrated Transport, 2003).

Summary

Labour came to power with high ambitions for policy on global warming and for policy on sustainable development more broadly. Institutions were reorganised or created to pursue this ambition and innovative policy was announced. However, the government was soon caught by the political realities which economic interests

---

Box 10.2 *The 2000 pre-budget statement: main points on green taxation for transport*

- A freeze in all road fuel and other oil duties – a cut in real terms in the price of petrol and diesel of 1½p per litre.
- An extension of the 'small' car threshold for vehicle excise duty from 1,200cc to 1,500cc, helping an additional 5.4 million car owners.
- Ultra-low-sulphur diesel cut by 3p per litre. This effectively meant a straightforward cut in the price of diesel because all diesel sold in Britain is ultra-low-sulphur diesel. This cost around £615 million in 2001/2.
- A 50 per cent cut in, and reform of, vehicle excise duty for lorries – and as a first step in this reform up to £265 million to be available for rebates on vehicle excise duty fees for the financial year.
- Abolition of vehicle excise duty on tractors and other agricultural vehicles.
- Cut in fuel duties in exchange for a road-user charge on all (including foreign) lorries in Britain.
- A £100 million 'modernisation fund' created to encourage hauliers to switch to cleaner vehicles by offering allowances for scrapping older lorries.

Source: Treasury press notice, 8 November 2000.

---

gave rise to and had to make some retreats. The majoritarian institutions of Britain allow political leaders to take effective leadership in setting out ambitious policy; however, the pluralist decision-making style also means policy is vulnerable to pressures from interested parties throughout policy-making process. The case of fuel duties, in particular, shows the difficulties government can have in defending the interests of future generations. The Chancellor had been emphasising how the fuel duty escalator was important in reducing carbon dioxide emissions. Few environmental arguments were made, however, for the need for fuel duties during the fuel crisis, and this again indicates the vulnerability of the interests of future generations in the face of the interests of present generations and electorates.

Nevertheless, global warming policy under Labour is clearly different from that under the Conservatives. Labour is more enthusiastic about steering the market towards a 'low-carbon' economy. This naturally raises a question: is the non-interventionist approach characterised by the Conservative government more to do with party ideology than with deep-seated state tradition?

An analysis of policy developments on renewable energy under Labour would help to answer this question. In 2002 the Labour government replaced the NFFO with the 'Renewables Obligation' as the principal tool to promote renewable energy. The Renewables Obligation requires electricity suppliers to obtain a specific proportion of electricity from renewable sources, to reach 10 per cent by 2010. Suppliers would meet the requirements by presenting Ofgem (the Office of Gas and Electricity Markets, the industry regulator) with Renewables Obligation certificates, which are tradable from generator to suppliers and between suppliers. Suppliers can bank the certificates, within limits, though not borrow them.

From the outset, there were concerns about its effectiveness. One of the reasons was that the government ruled out the incorporation of the provision of 'technology bands' in the Renewables Obligation. Without banding, fledgling technologies with large potential such as offshore wind and energy crops typically lose out against more established technologies. The government's view was that:

> it is no longer Government's job to pick winners.... The Government does not want to segment or unduly distort the marketplace.... Instead, it believes that competitive forces should be the drivers that shape the industry that emerges as a result of the introduction of the Obligation. (DTI, 2000b: 7)

The House of Commons Environment Audit Committee (2002) argued that while the government feared to be seen 'picking winners', it did so in any case by deciding which renewable technologies were eligible for the Renewables Obligation. Moreover, the government had introduced new funding for different renewable technologies. Nevertheless, the way the government provided funding was *ad hoc* and lacked strategic view. The government's desire to avoid the accusation of 'picking winners' and the contradictory desire to steer the market towards a low-carbon economy are perhaps underlying reasons for the lack of coherence in the government's funding policy and the establishment of new funding bodies, one after another, which were criticised by House of Commons Select Committee on Science and Technology (2003: para. 41). In nutshell, Britain lacks the capacity to mount a long-term, strategic intervention.

Labour was also reluctant to intervene in the newly established energy market system, the New Electricity Trading Arrangements (NETA) introduced in 2001. It was widely believed that the NETA represented a major barrier to the development of renewable sources of energy. However, both the government and Ofgem felt uneasy about making major revisions to the NETA. Ofgem was committed to enhancing competition and consumers' interests, as this was its primary statutory duty. Taking action to promote renewable energy would involve value judgements on trade-offs between consumer interests and environmental goals, which Ofgem thought was beyond its responsibilities (Ofgem, 2001). The government was also reluctant to intervene in free energy markets, and instead expected Ofgem to do so. As a result, important environmental decisions were left pending, having fallen somewhere between the government and Ofgem – a problem very similar to the EST's financial crisis under the Conservative government (see Chapter 7).

The Energy Act enacted in July 2004, however, introduced two notable interventionist provisions. The first was the introduction of a new duty on Ofgem to carry out its functions in the manner best calculated to 'contribute to the achievement of sustainable development'. An explicit environmental or sustainability duty on the regulator had long been called for, in fact since the Conservative era. Labour had been resistant to this call, but the 'surprising concession' came at the last stage of the passage of the bill (Collins, 2004). It is, however, a secondary duty, and what changes this will actually bring about remains to be seen. The second and more controversial provision is the government's concession to subsidise electricity distribution costs in the north of Scotland upon the introduction of British Electricity Trading and Transmission Arrangements, which essentially extend the NETA into Scotland. This move was strongly criticised by Ofgem as 'unnecessary and misguided' market intervention (Ofgem press release, 13 February 2004).

These changes in Labour's stances were perhaps influenced by growing criticism about the gap between government's professed ambition and actual delivery. The government was also under time pressure on the setting up of the British Electricity Trading and Transmission Arrangements (*ENDS Report* 352, 2004: 35). Still, these last-minute concessions indicate that Labour is more receptive, if reluctantly, to a more interventionist approach. Nevertheless, Labour's reluctance to take such steps and its aversion to being seen 'picking winners' suggest that non-interventionism is an underlying British policy style, and has constrained government climate policy.

Conclusions

The adoption of the Kyoto Protocol triggered renewed policy developments in both Japan and Britain. In Britain the change of government was another important opportunity for radical policy change. The patterns of policy developments were very different between them. Majoritarian institutions combined with ambitious politicians enabled Britain to make a step change. In Japan, its consensus policy styles delayed the introduction of GAT and policy developments were often the outcome of the pre-emption strategy prompted by new environmental and other challenges.

The cases of the Japanese GAT and the British CCL and domestic emissions trading scheme indicate the importance of policy sequences. To the extent that global warming policy is essentially about policy integration, the existing policies and the implications of global warming policy for them are necessarily crucial issues affecting a country's opportunities and constraints for developments on global warming policy. Policy developments are heavily 'path dependent'.

In both countries, economic interests had a crucial role in shaping policy. Different institutions in Japan and Britain affected the patterns of political processes, but when conflicts arose between economic interests and environmental interests, the former typically won over the latter. This, however, did not mean that policy contents were similar in the two countries. As seen in the cases of emissions trading, industries in Japan and Britain had different interests. While Japanese industry was against emissions trading, British industry lobbied for it. Industry's perceptions of what is in its interest were influenced by different institutions within which it operated. The corporatist industrial organisation provided Japanese industry with an institutional capacity for industry-led voluntary action with little government interference. Industry perceived this approach as preferable to, for example, a carbon tax and emissions trading. In Britain, on the other hand, lacking such institutional capacity, industry preferred more flexible emissions trading over the emerging CCL. Moreover, by taking the lead in developing a domestic emissions trading scheme, industry could effectively act as a co-decision-maker, enhancing its influence on policy. Institutions, thus, affected the preferences of industry, or more specifically industry's chosen strategy to reduce the climate policy burden on it, and this in turn affected global warming policy in Japan and Britain.

The question that arises then is whether this indicates a divergence in the stringency of global warming policy in Japan and Britain. On the face of it, Britain introduced both an energy tax on business and the domestic emissions trading scheme, which made Britain a world leader in climate politics, while Japan is now one of few major industrialised countries in which industry is still successful in rejecting the introduction of a carbon/energy tax. This is, however, too hasty an analysis. First, Japan is also a world leader in its innovative 'top-runner' approach to setting energy efficiency standards for electrical appliances and cars. Moreover, MITI's reform of energy-related taxation involved the introduction of taxes on carbon dioxide emissions. The question to be asked is how government is capable of implementing a policy of 'loss imposition' (Pierson and Weaver, 1993), which is inherent in emissions policy. The fundamental picture that arises is, then, that the most affected groups were effective in winning the government's favour. By arguing this, I do not mean that new policies and measures introduced have increased policy stringency very little. The CCL was, for example, certainly effective in compelling energy-intensive industries to commit themselves to taking action through negotiated agreements, which the government had failed to make them accept. Policy developments since the Kyoto conference did make global warming policy in both Japan and Britain more stringent than before. Nevertheless, the policy developments were, after all, very much influenced by the interests of the most affected. The British experiences on all three cases reinforce the main findings of this study: the fundamental constraint on global warming policy stems from issue characteristics, and economic interests crucially affect the important aspects of policy, regardless of institutions.

While the Japanese and British approaches to global warming policy are different – policy in the former has typically featured voluntary action by industry and high expectations for environmental technologies such as LEVs, and policy in the latter has centred on increasing use of economic instruments – a trend towards convergence can detected. For example, with the introduction of CCLAs in Britain, both countries now have a formal system of voluntary action by industry. Japan and Britain also have a similar policy on the promotion renewable energy, with the former emulating the British approach. More generally, Britain now puts more emphasis on 'processes' of policy, apparently adopting the corporatist style of consensus and collaboration, as seen in long consultation exercises before the introduction of major policies.

What are the implications in theoretical terms? A clear implication is that policy-makers are in a ceaseless quest for workable policy. Policy-makers 'puzzle' over what they can do in the face of a set of constraints arising from issue characteristics. Confronted with similar problems, they look at ideas and experiences in other countries, to learn from and emulate them. Given that policy-makers will look for rational and feasible policies, the issue-based account helps explain a general trend towards similar policy approaches. This issue-based account, however, needs an institutionalist corrective. The clearly different nature of voluntary action in Japan and Britain is a consequence of their unique institutional opportunities and constraints. More generally, policy emulation is affected by domestic political conjunctures. The Japanese choice of the British, market-based approach to the promotion of renewable energy is the result of various combined factors, including the initial political move to introduce a German-style fixed-price system, the power companies' opposition to it and MITI's move to forestall it and search for a system compatible with the ongoing electricity deregulation and liberalisation. The British efforts to introduce economic instruments to tackle carbon dioxide emissions were also considerably affected by the fuel poverty problem.

With policy-makers searching for measures which are more cost-effective, politically feasible and compatible with the developing liberal international trade regime, a general trend towards convergence in policy will continue and is likely to be accelerated within the international climate regime – specifically, the Kyoto Protocol, which includes specific provisions to meet Kyoto targets. This will provide a rich source for further work on comparative analysis of climate policy. Similarities and differences in the details of policy would highlight underlying opportunities and constraints within a country's policy processes.

Notes

1 For example, MITI started to use the Energy Conservation Law to facilitate industry's voluntary action. Factory checks based on the Law are prioritised in order to target industries that have not established voluntary plans or those whose progress lags behind the targets in their plans.

2 The baseline year for sulphur hexafluoride, hydrofluorocarbons and perfluorocarbons is 1995, while that for carbon dioxide, nitrous oxide and methane is 1990.

3 This includes small and medium-sized hydroelectric power plants.

4 The Climate Change Programme envisaged that emissions trading would deliver reductions of 7.3 million tonnes of carbon dioxide (or 2 million tonnes of carbon) by 2010, and the domestic emissions trading scheme was expected to deliver about 1.8 million tonnes of carbon dioxide (0.5 million tonnes of carbon). The remaining 5.5 million tonnes of carbon dioxide (1.5 million tonnes of carbon) was to come from the EU emissions trading scheme.

5 The German plan, for example, allowed industry to emit more carbon dioxide than the existing voluntary agreement between industry and government would suggest.

6 Although the government did not change the total number of allocations, because the DTI increased its projections for emissions of carbon dioxide, the projected reduction by 2010, including the contribution from the EU trading scheme, fell.

7 The CBI's director, Digby Jones, together with Secretary of the State at DEFRA, issued a joint statement on the matter to the Commission. See *ENDS Report* (353, 2004: 39–40).

8 The Renewables Obligation, which replaced the NFFO under Labour, put an extra 10 per cent on household electricity bills. However, this was not aimed at promoting energy efficiency in the household sector.

Also there was a strong concern about its effects on the poor and the government carefully limited price increases within the range of insignificant.

9 The Conservatives went on to pledge to cut fuel duty by 6p per litre in its election manifesto.

References

Advisory Committee on Business and the Environment (1998) *Climate Change: A Strategic Issue for Business*, London: Department of the Environment, Transport and the Regions.

Andersen, M. S. and D. Liefferink (eds) (1997) *European Environmental Policy: The Pioneers*, Manchester: Manchester University Press.

ANRE (1990) *Enerugî: Shinchôryû eno Chôsen* [Energy: the challenge to the new trend], interim report by the Advisory Committee for Energy, Tokyo: International Trade and Industry Research Centre.

ANRE (1993) *Enerugî Seisaku no Ayumi to Tenbo* [Energy policy in the past and future], Tokyo: International Trade and Industry Research Centre.

ANRE (1994a) *Enerugî: Shinseiki eno Shinario* [Energy: the scenario for the new era], interim report by the Supply and Demand Subcommittee, Advisory Committee for Energy, Tokyo: International Trade and Industry Research Centre.

ANRE (1994b) *Shinenerugî Dônyû Taikô* [The basic guidelines for new energy introduction], Tokyo: MITI.

ANRE (1997) *Enerugî: Mirai Karano Keisyô* [Energy: tasks for future energy policies], interim report by the Subcommittee on Basic Policy Directions, Advisory Committee for Energy, Tokyo: International Trade and Industry Research Centre.

ANRE (2001) *Mitsumeyô! Wagakuni no Enerugî* [Our country's energy], Tokyo: International Trade and Industry Research Centre.

ANRE (2003) *Energy in Japan 2003*, Tokyo: Communications Office, Agency for Natural Resources and Energy, Ministry of International Trade and Industry, www.enecho.meti.go.jp (last accessed 14 June 2005).

Arts, B. and W. Rüdig (1995) Negotiating the 'Berlin Mandate': reflections on the first 'Conference of the Parties' to the UN Framework Convention on Climate Change, *Environmental Politics* 4, 481–7.

Ayukawa, Y. and WWF (2001) Nihon ha sugunidemo hijyun ha dekiru [Japan can ratify the protocol immediately], www.wwf.or.jp/lib/climate/repcope.htm (last accessed 14 June 2005)

Barret, S. (1991) Economic instruments for climate change policy, pp. 51–108 in: OECD, *Responding to Climate Change: Selected Economic Issues*, Paris: OECD.

Beuermann, C. (2000) Germany: regulation and the precautionary principle, pp. 85–111 in: Lafferty, W. M. and J. Meadowcroft (eds), *Implementing Sustainable Development: Strategies and Initiatives in High Consumption Societies*, Oxford: Oxford University Press.

Beuermann, C. and J. Jäger (1996) Climate change politics in Germany: how long will any double dividend last?, pp. 186–227 in: O'Riordan, T. and J. Jäger (eds), *Politics of Climate Change: A European Perspective*, London: Routledge.

Blyth, M. (2002) Institutions and ideas, pp. 292–310 in: Marsh, D. and G. Stoker (eds), *Theory and Methods in Political Science*, Basingstoke: Palgrave Macmillan.

Boehmer-Christiansen, S. A. (1995) Britain and the Intergovernmental Panel on Climate Change: the impacts of scientific advice on global warming. Part 2: The domestic story of the British response to climate change, *Environmental Politics* 4, 175–96.

Boehmer-Christiansen, S. A. and J. Skea (1991) *Acid Politics: Environmental and Energy Policies in Britain and Germany*, London: Belhaven Press.

Boyd, R. (1987) Government–industry relations in Japan: access, communication, and competitive collaboration, pp. 61–89 in: Wilks, S. and M. Wright (eds), *Comparative Government–Industry Relations: Western Europe, the United States, and Japan*, Oxford: Clarendon Press.

Bressers, H. T. A. and L. J. O'Toole, Jr (1998) The selection of policy instruments: a network-based perspective, *Journal of Public Policy* 18, 213–39.

British Government Panel on Sustainable Development (1997) *Third Report*, www.sd-commission.org.uk/panel-sd/panel3/index.htm (last accessed 31 July 2005).

Broadbent, J. (1996) *Environmental Politics in Japan: Networks of Power and Protest*, Cambridge: Cambridge University Press.

Brown, M. (1994) Combined heat and power: positive progress in the UK, *Energy Policy* 22, 173–7.

Calder, K. E. (1988) *Crisis and Compensation: Public Policy and Political Stability in Japan*, Princeton, NJ: Princeton University Press.

Cavender-Bares, J. and J. Jäger, with R. Ell (2001) Global environmental risk management in Germany, pp. 61–92 in: Social Learning Group, *Learning to Manage Global Environmental Risks. Vol. 1: A*

Comparative History of Social Responses to Climate Change, Ozone Depletion, and Acid Rain, Cambridge, MA: MIT Press.

CBI (1989) *The Capital at Risk: Transport in London Task Force Report*, London: CBI.

Climate Change Secretariat (2002) *A Guide to the Climate Change Convention and Its Kyoto Protocol*, Bonn: UNFCCC, http://unfccc. int/resource/guideconvkp-p.pdf (last accessed 14 June 2005).

Climate Network Europe and United States Climate Action Network (1996) *Independent NGO Evaluation of National Plans for Climate Change Mitigation: OECD Countries (Draft)*, unpublished paper.

Collier, U. (1997) 'Windfall' emission reductions in the UK, pp. 87–107 in: Collier, U. and R. E. Löfstedt, *Cases in Climate Change Policy. Political Reality in the European Union*, London: Earthscan.

Collins, J. (Green Alliance) (2004) Green lights for micro-renewable energy, *Parliamentary Newsletter*, 27 August–10 September.

Commission for Integrated Transport (2003) *10 Year Transport Plan: Second Assessment Report*, www.cfit.gov.uk/reports/10year/second/ (last accessed 31 July 2005).

Council for the Coal Industry (1999) *Genkô no Sekitan Seisaku no Enkatsu na Kanryô ni muketeno Susumekata ni tsuite* [The way to smoothly terminate the current coal policy], Tokyo: MITI.

Council of Cabinet Ministers for the Promotion of Comprehensive Energy Policy (1997) *2000 nen ni muketa Sôgôteki na Shô-enerugî Taisaku* [Comprehensive energy conservation policy towards 2000], 1 April, www.eccj.or.jp/2000/2000-2j.html (last accessed 31 July 2005).

Cox, G. W. and M. F. Thies (1998) The cost of intraparty competition: the single, nontransferable vote and money politics in Japan, *Comparative Political Studies* 31, 267–91.

Crepaz, M. M. L. (1995) Explaining national variations of air pollution levels: political institutions and their impact on environmental policy-making, *Environmental Politics* 3, 391–414.

Crepaz, M. M. L. and A. Lijphart (1995) Linking and integrating corporatism and consensus democracy: theory, concepts and evidence, *British Journal of Political Science* 25, 281–8.

Dahl, R. (1961) *Who Governs?*, New Haven, CN: Yale University Press.

Daugbjerg, C. (1998) Linking policy networks and environmental policies: nitrate policy making in Denmark and Sweden 1970–1995, *Public Administration* 76, 275–94.

Daugbjerg, C. and G. T. Svendsen (2001) *Green Taxation in Question: Politics and Economic Efficiency in Environmental Regulation*, Basingstoke: Palgrave.

David Davies Associates (1996) *At the Crossroads: Investing in Sustainable Local Transport. A Review of Funding for Local Transport and the 1996–97 TPP Submissions*, London: Transport 2000 and CPRE.

De Shalit, A. (1995) Is liberalism environment-friendly?, *Social Theory and Practice* 21, 287–314.

DEFRA (2004a) *e-Digest of Environmental Statistics*, www.defra.gov.uk/environment/statistics/index.htm (last accessed 14 June 2005).

DEFRA (2004b) *EU Emissions Trading Scheme. UK Draft National Allocation Plan. 2005–2007*, January, www.defra.gov.uk/corporate/consult/eu-etsnap/index.htm (last accessed 14 June 2005).

DEFRA (2004c) *EU Emissions Trading Scheme. UK National Allocation Plan. 2005–2007*, May, www.defra.gov.uk/corporate/consult/euetsnapstagethree/index.htm (last accessed 14 June 2005).

DEn (1983) *Investment in Energy Use as an Alternative to Investment in Energy Supply*, London: HMSO.

Desai, U. (ed.) (2002) *Environmental Politics and Policy in Industrialised Countries*, Cambridge, MA: MIT Press.

DETR (2000a) *Climate Change: Draft UK Programme*, www.defra.gov.uk/environment/climatechange/draft/index.htm (last accessed 14 June 2005).

DETR (2000b) *Climate Change: The UK Programme*, Cm 4913, London: TSO.

DOE (1992) *Climate Change: Our National Programme for CO$_2$ Emissions. A Discussion Document*, London: DoE.

DOE (1993) *Climate Change: Our National Programme for CO$_2$ Emissions. Report of Conference Held at QEII Conference Centre on 7 May 1993*, London: DoE.

DOE (1995) *Climate Change: The UK Programme. Progress Report on Carbon Dioxide Emissions*, London: DoE.

Doherty, B. (1997) Changing environmentalism in the 1990s, www.psa.ac.uk/cps/1997%5Cdohe.pdf (last accessed 14 June 2005).

Domestic Institutions Subcommittee, Global Environment Committee, Central Environment Council (2001) *Chûkan Torimatome (An)* [The interim report (Draft)], www.env.go.jp/council/06earth/r061-01/01.pdf (last accessed 31 July 2005).

DOT (1989a) *National Road Traffic Forecasts (Great Britain) 1989*, London: HMSO.

DOT (1989b) *Roads for Prosperity*, Cm 693, London: HMSO.

DOT (1996) *Transport – The Way Forward: The Government Response to the Transport Debate*, Cm 3234, London: HMSO.

DOT (2004) *Transport Statistics for Great Britain* (2004 edn), www.dft.gov.uk/stellent/groups/dft_transstats/documents/page/dft_transstats_031999.hcsp (last accessed 14 June 2005).

Downs, A. (1972) Up and down with ecology – the 'issue attention cycle', *Public Interest* 28, 38–50.

Dryzek, J. S. (1997) *The Politics of the Earth: Environmental Discourses*, Oxford: Oxford University Press.

DTI (1992) *Energy Related Carbon Emissions in Possible Future Scenarios for the United Kingdom*, Energy Paper 59, London: HMSO.

DTI (1993) *The Prospects for Coal: Conclusions of the Government's Coal Review*, Cm 2235, London: HMSO.

DTI (1995a) *Energy Projections for the UK: Energy Use and Energy-Related Emissions of Carbon Dioxide in the UK, 1995–2020*, Energy Paper 65, London: HMSO.

DTI (1995b) *Energy Report 1995: Competition, Competitiveness, and Sustainability*, London: HMSO.

DTI (2000a) *Energy Report 2000: Market Reforms and Innovation*, London: HMSO.

DTI (2000b) *New and Renewable Energy. Prospects for the 21st Century. The Renewable Obligation. Preliminary Consultation*, www.dti.gov.uk/energy/renewables/publications/pdfs/ropc.pdf (last accessed 14 June 2005).

DTI (2001) *Energy Trends Special Features and Articles: January 2000 to December 2001*, December, www.dti.gov.uk/energy/inform/energy_trends/features.shtml (last accessed 3 August 2005).

DTI (2003) *Our Energy Future – Creating a Low Carbon Economy*, Cm 5761, London: TSO.

DTI (2004) *Digest of Energy Statistics*, www.dti.gov.uk/energy/inform/dukes/index.shtml (last accessed 14 June 2005).

Dudley, G. and J. Richardson (2000) *Why Does Policy Change? Lessons from British Transport Policy 1945–99*, London: Routledge.

Duncan, M. (2004) *Contemporary Japan* (2nd edn), New York: Palgrave Macmillan.

Dunleavy, P. and B. O'Leary (1987) *Theories of the State: The Politics of Liberal Democracy*, Basingstoke: Macmillan.

Dunleavy, P., A. Gamble, R. Heffernan and G. Peel (eds) (2003) *Developments in British Politics 7*, Basingstoke: Palgrave Macmillan.

EA (1990a) *Chikyûondanka Bôshi Kôdô Keikaku* [Action programme to arrest global warming], Tokyo: EA.

EA (1990b) *Kankyô Hakusyo. Sôsetsu* [White paper on the environment], Tokyo: Printing Bureau, Ministry of Finance.

EA (1990c) *Nisankatanso Haisyutsu Anteika Seisaku Opushon ni Kansuru Kêsu Sutadî* [Case study on policy options for the stabilisation of CO_2 emissions], Tokyo: EA.

EA (1994a) *The Basic Environmental Plan*, Tokyo: Printing Bureau, Ministry of Finance.

EA (1994b) *Quality of the Environment in Japan*, Tokyo: Printing Bureau, Ministry of Finance.

EA (1994c) *Teikôgaisha fukyû hôsaku kentôkai saishûhôkoku* [The final report by the study group to examine policy to promote low-emission vehicles], Tokyo: EA.

EA (1996) *Nihon no Kankyô Taisaku wa Susunde Iruka: Kankyô kihon keikaku no dai-ikkai tenken hôkoku* [Has Japanese environmental policy made progress? The first progress report on the Basic Environmental Plan by the Central Environment Council], Tokyo: Printing Bureau, Ministry of Finance.

EA (1997a) *Nihon no Kankyô Taisaku wa Susunde Iruka II: Kankyô kihon keikaku no dai-nikai tenken hôkoku* [Has Japanese environmental policy made progress? The second progress report on the Basic Environmental Plan by the Central Environment Council], Tokyo: Printing Bureau, Ministry of Finance.

EA (1997b) *Kankyô Hakusyo 1997* [White paper on the environment], Tokyo: Printing Bureau, Ministry of Finance.

EA (1998) *Nihon no Kankyô Taisaku wa Susunde Iruka III: Kankyô kihon keikaku no dai-sankai tenken hôkoku* [Has Japanese environmental policy made progress? The third progress report on the Basic Environmental Plan by the Central Environment Council], Tokyo: Printing Bureau, Ministry of Finance.

Eccleston, B. (1995) The influence of Japanese NGOs on environmental policy making, *Hitotsubashi Journal of Social Studies* 27, 145–56.

Economic Planning Agency (1992) *Seikatsu Taikoku 5 kanen Keikaku – Chikyûshakai tono Kyôzon o Mezashite* [Five-Year Plan for Livelihood Great Power – sharing a better quality of life around the globe], Tokyo: Printing Bureau, Ministry of Finance.

Elder, M. (2002) The Japanese challenge to conventional (Western) theories of corporate trade policy preferences, Prepared for delivery at the Annual Convention of the International Studies Association in New Orleans, Louisiana, 24–27 March.

Energy Conservation Centre (2003) *Handbook of Energy and Economic Statistics in Japan 2003*, Tokyo: Energy Conservation Centre.

Energy Information Administration (1996) *Renewable Energy Annual 1996*, www.eia.doe.gov/cneaf/solar.renewables/renewable.energy. annual/contents.html (last accessed 16 July 2005).

Enloe, C. H. (1975) *The Politics of Pollution in a Comparative Perspective: Ecology and Power in Four Nations*, New York: David McKay.

ETSU (2001) *Climate Change Agreements – Sectoral Energy Efficiency Targets*, Harwell: AEA Technology.

Ezawa, M. (1998) *Dare ga Kankyô Hozen-hi wo Futan Surunoka* [Who bears the burden of environmental protection?], Tokyo: Cyuo Keizaisha.

Fermann, G. (1992) *Japan in the Greenhouse: Responsibilities, Policies and Prospects for Combating Global Warming*, EED report 13, Lysaker: Fridtjof Nansen Institute.

FOE (2003) *02/03 Review: Join in Our Successes*, London: FOE.

FOE (2004) Blair caves into industry on climate, press release, 27 October, www.foe.co.uk/resource/press_releases/blair_caves_into_industry_27102004.html (last accessed 14 June 2005).

Freeman, G. P. (1985) National style and policy sectors: explaining structural variation, *Journal of Public Policy* 5, 467–96.

Fukushima, M. (1995) Syô enerugî seisaku [Energy efficiency policy], pp. 258–75 in: Matsui, K. (ed.), *Enerugî Sengo 50 nen no Kensyô* [Fifty years of energy policy after the war], Tokyo: Denryokushimpôsya.

Gay, C., T. Palvolgyi, E. Evans, J. Corfee-Morlot, R. Hornung and K. Simeonova (1999) *Report on the In-depth Review of the Second National Communication of the United States of America*, FCCC/IDR.2/USA, Bonn: UNFCCC, http://unfccc.int/resource/docs/idr/usa02.htm (last accessed 14 June 2005).

Goodwin, P. (1999) Transformation of transport policy in Great Britain, *Transportation Research Part A* 33, 655–69.

Government of Japan (1998) Shin Dôro Keikaku 5 kanen Keikaku [New 5-year plan for road improvement], 29 May, www.mlit.go.jp/road/consider2/keikaku/index1.html (last accessed 16 July 2005).

Gresser, J., K. Fujikura and A. Morishima (1981) *Environmental Law in Japan*, Tokyo: MIT Press.

Haas, P. M. (1989) Do regimes matter? Epistemic communities and Mediterranean pollution control, *International Organization* 43, 377–403.

Haas, P. M. (1990) Obtaining international environmental protection through epistemic consensus, *Millennium* 19, 347–63.

Haas, P. M. (1992) Introduction: epistemic communities and international policy co-ordination, *International Organisation* 46, 1–35.

Haigh, N. (1996) Climate change policies and politics in the European Community, pp. 155–85 in: O'Riordan, T. and J. Jäger (eds), *Politics of Climate Change: A European Perspective*, London: Routledge.

Haigh, N. and C. Lanigan (1995) Impact of the European Union on UK environmental policy making, pp. 18–37 in: Gray, T. S. (ed.), *UK Environmental Policy in the 1990s*, London: Macmillan.

Hajer, M. A. (1995) *The Politics of Environmental Discourse: Ecological Modernization and the Policy Process*, Oxford: Clarendon Press.

Haley, J. O. (1995) Japan's postwar civil service: the legal framework, pp. 77–101 in: Kim, H., M. Muramatsu, T. J. Pempel and K. Yamamura (eds), *The Japanese Civil Service and Economic Development*, Oxford: Clarendon Press.

Hall, P. A. (1986) *Governing the Economy: The Politics of State Intervention in Britain and France*, New York: Oxford University Press.

Hall, P. A. (ed.) (1989) *The Political Power of Economic Ideas: Keynesianism Across Nations*, Princeton, NJ: Princeton University Press.

Hall, P. A. (1992) The movement from Keynesianism to monetarism: institutional analysis and British economic policy in the 1970s, pp. 90–113 in: Steinmo, S., K. Thelen and F. Longstreth (eds), *Structuring Politics: Historical Institutionalism in Comparative Analysis*, New York: Cambridge University Press.

Hall, P. A. (1993) Policy paradigm, social learning, and the state: the case of economic policymaking in Britain, *Comparative Politics* 25, 275–94.

Hall, P. A. (1995) The Japanese civil service and economic development in comparative perspectives, pp. 484–505 in: Kim, H., M. Muramatsu, T. J. Pempel and K. Yamamura (eds), *The Japanese Civil Service and Economic Development*, Oxford: Clarendon Press.

Hall, P. A. and R. C. R. Taylor (1996) Political science and the three new institutionalisms, *Political Studies* 44, 936–57.

Hall, P. A. and R. C. R. Taylor (1998) The potential of historical institutionalism: a response to Hay and Wincott, *Political Studies* 46, 958–62.

Harris, R. (ed.) (1997) *The Collected Speeches of Margaret Thatcher*, London: HarperCollins.

Harvey, K. (1994) The development of combined heat and power in the UK, *Energy Policy* 22, 179–81.

Hay, C. and D. Wincott (1998) Structure, agency and historical institutionalism, *Political Studies* 46, 951–7.

Hayashi, Y. and J. Nagamine (2001) Dôro tôshi no seisaku kettei purosesu [Decision-making process on road policy], pp. 22–40 in: Nagamine, J. and T. Katayama (eds), *Kôkyô-Tôshi to Dôro Seisaku* [Public works investment and road policy], Tokyo: Keiso Shobo.

Helm, D. (2003) *Energy, the State, and the Market. British Energy Policy Since 1979*, Oxford: Oxford University Press.

Henkel, H. (President of the Federation of German Industries) and S. Toyoda (Chairman of the Japan Federation of Economic Organisations) (1996) Joint statement on global warming prevention, 1 November, www.keidanren.or.jp/english/policy/pol049.html (last accessed 14 June 2005).

Her Majesty's Government (1990) *This Common Inheritance: Britain's Environmental Strategy*, Cm 1200, London: HMSO.

Her Majesty's Government (1992) *This Common Inheritance. The Second Year Report*, Cm 2068, London: HMSO.

Her Majesty's Government (1994a) *Climate Change: The UK Programme*, Cm 2427, London: HMSO.

Her Majesty's Government (1994b) *Sustainable Development: The UK Strategy*, Cm 2426, London: HMSO.

Her Majesty's Government (1996) *This Common Inheritance: UK Annual Report 1996*, Cm 3188, London: HMSO

Her Majesty's Government (1997a) *Civil Service Yearbook*, London: HMSO.

Her Majesty's Government (1997b) *Climate Change: The UK Programme. United Kingdom's Second Report under the Framework Convention on Climate Change*, Cm 3558, London: HMSO.

Hill, M. (1997) *The Policy Process in Modern State*, Hemel Hempstead: Prentice Hall/Harvester Wheatsheaf.

Houghton, J. (2002) Overview of the climate change issue, Forum 2002 on Climate Change, at St Anne's College, Oxford, 14–17 July.

Houghton, J. T., B. A. Callander and S. K. Varney (eds) (1992) *Climate Change 1992. The Supplementary Report to the IPCC Scientific Assessment*, Cambridge: Cambridge University Press.

Houghton, J. T., G. T. Jenkins and J. T. Ephraums (eds) (1990) *Climate Change: The IPCC Scientific Assessment*, Cambridge: Cambridge University Press.

House of Commons Energy Select Committee (1989) *Energy Policy Implications of Greenhouse Effect*, Session 1988–89, HC 192-I, London: HMSO.

House of Commons Energy Select Committee (1991) *Energy Efficiency*, Session 1990–91, HC 91-I, London: HMSO.

House of Commons Energy Select Committee (1992) *Renewable Energy*, Session 1991–92, HC 43-I, London: HMSO.

House of Commons Environment Audit Committee (1999a) *Energy Efficiency*, Session 1998–99, HC 571-i, London: TSO.

House of Commons Environment Audit Committee (1999b) *The Budget 1999: Environmental Implications*, Session 1998–99, HC 326, London: TSO.

House of Commons Environment Audit Committee (2002) *A Sustainable Energy Strategy? Renewables and the PIU Review*, Session 2001–02, HC 582-I, London: TSO.

House of Commons Environment Audit Committee (2003) *Energy White Paper – Empowering Change?*, Session 2002–03, HC 618, London: TSO.

House of Commons Environment Audit Committee (2004) *Budget 2004 and Energy*, Session 2003–04, HC 490, London: TSO.

House of Commons Environment Select Committee (1993) *The Environmental Implications of Energy Policy*, Session 1992–93, HC 465-i, London: HMSO.

House of Commons Environment, Transport and Regional Affairs Committee (2000a) *UK Climate Change Programme*, Session 1999–2000, HC 194-I, London: TSO.

House of Commons Environment, Transport and Regional Affairs Committee (2000b) *The Road Haulage Industry*, Session 1999–2000, London: TSO.

House of Commons Select Committee on Science and Technology (2003) *Towards a Non-carbon Fuel Economy: Research, Development and Demonstration*, Session 2002–03, HC 55-I, London: TSO.

House of Lords Select Committee on Science and Technology (1996) *Towards Zero Emissions for Road Transport*, Session 1996–97, HL 13, London: HMSO.

House of Lords Select Committee on Sustainable Development (1994) *Report from the Select Committee on Sustainable Development*, Session 1993–94, HL 42, London: HMSO.

House of Lords Select Committee on the European Communities (1992) *Carbon/Energy Tax*, Session 1991–92, London: HMSO.

Howlett, M. (1991) Policy instruments, policy styles, and policy implementation: national approaches to theories of instrument choice, *Policy Studies Journal* 19, 1–21.

Hukkinen, J. (1995a) Corporatism as an impediment to ecological sustenance: the case of Finnish waste management, *Ecological Economics* 15, 59–75.

Hukkinen, J. (1995b) Long-term environmental policy under corporatist institutions, *European Environment* 5, 98–105.

IEA (1993) *Energy Balances of OECD Countries*, Paris: IEA/OECD.

IEA (2002) *Energy Balances of OECD Countries*, Paris: IEA/OECD.

Igarashi, T. and A. Ogawa (eds) (1997) *Kôkyôjigyô o Dôsuruka* [What to do about public works projects], Tokyo: Iwanami Shinsyo.

Industrial Structure Council, Advisory Committee for Energy and Industrial Technology Council (1992) *Fourteen Proposals for New Earth – Policy Triad for the Environment, Economy and Energy*, Tokyo: MITI.

IPCC (1991) *Climate Change: The IPCC Response Strategies*, Washington, DC: Island Press.

IPCC (1994) *The 1994 Report of the Scientific Assessment Working Group of IPCC, Summary for Policymakers*, Geneva: WMO/UNEP.

IPCC (1996) *Climate Change 1995. IPCC Second Assessment*, www.ipcc. ch/pub/sa(E).pdf (last accessed 14 June 2005).

IPCC (2001a) *Climate Change 2001: The Scientific Basis*, www.grida.no/ climate/ipcc_tar/wg1/index.htm (last accessed 14 June 2005).

IPCC (2001b) *Climate Change 2001: Synthesis Report*, www.grida.no/ climate/ipcc_tar/vol4/english/index.htm (last accessed 14 June 2005).

IPCC (2001c) *Climate Change 2001; Mitigation*, www.grida.no/climate/ ipcc_tar/wg3/index.htm (last accessed 15 July 2005).

Jänicke, M. (1995) The political system's capacity for environmental policy, *Forschungsstelle für Umweltpolitik*, report no. 95–6, Berlin.

Jänicke, M. (1997) The political system's capacity for environmental policy, pp. 1–24 in: Jänicke, M. and H. Weidner (eds), *National Environmental Policies: A Comparative Study of Capacity-Building*, Berlin: Springer.

Jänicke, M. and H. Weidner (eds) (1995) *Successful Environment Policy: A Critical Evaluation of 24 Cases*, Berlin: Edition Sigma.

Japan Environment Council (1994) *Kankyô Kihonhô o Kangaeru* [Considering the Basic Environment Law], Tokyo: Jikkyô Syuppan.

Japan Federation of Bar Associations (1999) *Koritsu suru Nihon no Enerugî Seisaku: Enerugî Seisaku ni kansuru Chôsa Hôkoku* [Isolated Japanese energy policy: report on energy policy], Tokyo: Nanatsumorisyokan.

Jenkins-Smith, H. C. and P. A. Sabatier (1993) The dynamics of policy-oriented learning, pp. 41–56 in: Sabatier, P. and H. Jenkins-Smith (eds), *Policy Change and Learning: An Advocacy Coalition Approach*, Boulder, CO: Westview Press.

Johnson, C. (1982) *MITI and the Japanese Miracle, 1925–1975*, Tokyo: Charles E. Tuttle.

Joint Meeting of Councils Relating to Domestic Measures to Arrest Global Warming (1997) The basic direction of global warming policy centring comprehensive measures to restrain energy demand, November, www.kantei.go.jp/jp/ondan/1128houkoku.html (last accessed 1 August 2005).

Jordan, A. (2002) *The Europeanization of British Environmental Policy. A Departmental Perspective*, Basingstoke: Palgrave Macmillan.

Jordan, A. G. and T. O'Riordan (1993) Sustainable development: the political and institutional challenge, pp. 183–202 in: Pearce, D. (ed.), *Blueprint 3: Measuring Sustainable Development*, London: Earthscan.

Jordan, A. G. and J. Richardson (1982) The British policy style or the logic of negotiation?, pp. 80–110 in: Richardson, J. (ed.), *Policy Styles in Western Europe*, London: George Allen and Unwin.

Jordan, A. G. and J. Richardson (1987) *Government and Pressure Groups in Britain*, Oxford: Clarendon Press.

Kamei, Y. (2002) *Nihon Enerugî Kaizô-ron* [Energy reform in Japan], Tokyo: Energy Forum.

Kanemoto, Y., K. Hasuike and T. Fujiwara (2003) Road transport and environmental policies in Japan, RIETI Discussion Paper Series, 01-E-003, Tokyo: Research Institute of Economy, Trade and Industry, www.rieti.go.jp/jp/publications/dp/01e003.pdf (last accessed 31 July 2005).

Kasa, S. (1999) *Social and Political Barriers to Green Tax Reform: The Case of CO_2 Taxes in Norway*, CICERO Policy Note 5, Oslo: Center for International Climate and Environmental Research, www.cicero.uio.no/media/57.pdf (last accessed 31 July 2005).

Kashima, S. (2003) *Chikyû Kankyô Seiki no Jidôsha Zeisei* [Vehicle taxation in the global environment century], Tokyo: Keiso Shobo.

Katzenstein, P. J. (1978) Conclusion: domestic structures and strategies of foreign economic policy, pp. 295–336 in: Katzenstein, P. J. (ed.),

Between Power and Plenty: Foreign Economic Policies of Advanced Industrial States, Madison, WI: University of Wisconsin Press.

Katzenstein, P. J. (1985) *Small States in World Markets: Industrial Policy in Europe*, Ithaca, NY: Cornell University Press.

Kavanagh, D. (1990) *British Politics: Continuities and Change*, Oxford: Oxford University Press.

Keidanren (1989) Chikyû kankyô mondai ni taisuru waga kuni sangyôkai no torikumi no genjyô to kongo no hôkô [Japanese industry's policy on global environmental problems], October, Tokyo: Keidanren.

Keidanren (1990) Basic opinions on global environmental problems, April, Tokyo: Keidanren.

Keidanren (1991) Keidanren Global Environment Charter, www. keidanren.or.jp/english/speech/spe001/s01001/s01b.html (last accessed 14 June 2005).

Keidanren (1996) Keidanren Appeal on the Environment: declaration on voluntary action of Japanese industry directed at conservation of global environment in the 21st century, 16 July, www.keidanren.or.jp/ english/policy/index07.html (last accessed 14 June 2005).

Keman, H. and P. Pennings (1995) Managing political and societal conflict in democracies: do consensus and corporatism matter?, *British Journal of Political Science* 25, 271–81.

Keohane, R. O., P. M. Hass and M. A. Levy (1993) The effectiveness of international environmental institutions, pp. 3–24 in: Haas, P. M., R. O. Keohane and M. A. Levy (eds), *Institutions for the Earth: Sources of Effective International Environmental Protection*, Cambridge, MA: MIT Press.

King, D. S. (1992) The establishment of work-welfare programs in the United States and Britain: politics, ideas and institutions, pp. 217–50 in: Steinmo, S., K. Thelen and F. Longstreth (eds), *Structuring Politics: Historical Institutionalism in Comparative Analysis*, New York: Cambridge University Press.

Kingdon, J. (1984) *Agenda, Alternatives, and Public Policies*, Boston: Little, Brown.

Krasner, S. (1983) Structural causes and regime consequences: regimes as intervening variables, in: Krasner, S. (ed.), *International Regimes*, Ithaca, NY: Cornell University Press.

Krasner, S. D. (1984) Approach to the state: alternative conceptions and historical dynamics, *Comparative Politics* 16, 223–46.

Kronsell, A. (1997) Sweden: setting a good example, pp. 40–80 in: Andersen, M. S. and D. Liefferink (eds), *European Environmental Policy. The Pioneers*, Manchester: Manchester University Press.

Lafferty, W. M. and E. Hovden (2003) Environmental policy integration: towards an analytical framework, *Environmental Politics* 12:3, 1–22.

Lafferty, W. M. and J. Meadowcroft (eds) (2000a) *Implementing*

Sustainable Development: Strategies and Initiatives in High Consumption Societies, Oxford: Oxford University Press.

Lafferty, W. M. and J. Meadowcroft (2000b) Concluding perspectives, pp. 422–59 in: Lafferty, W. M. and J. Meadowcroft (eds), *Implementing Sustainable Development: Strategies and Initiatives in High Consumption Societies*, Oxford: Oxford University Press.

Lamont, N. (1999) *In Office*, London: Little, Brown.

Langhelle, O. (2000) Norway: reluctantly carrying the torch, pp. 174–208 in: Lafferty, W. M. and J. Meadowcroft (eds), *Implementing Sustainable Development: Strategies and Initiatives in High Consumption Societies*, Oxford: Oxford University Press.

Latham, E. (1964) The group basis of politics: notes for theory, pp. 32–57 in: Munger, F. and D. Price (eds), *Political Parties and Pressure Groups*, New York: Crowell.

Lehmbruch, G. and P. C. Schmitter (eds) (1982) *Patterns of Corporatist Policy-Making*, New York: Sage.

Lesbirel, S. H. (1991) Structural adjustment in Japan: terminating 'Old King Coal', *Asian Survey* 31, 1079–94.

Lijphart, A. (1984) *Democracies: Patterns of Majoritarian and Consensus Government in Twenty-One Countries*, New Haven, CT: Yale University Press.

Lijphart, A. (1999) *Patterns of Democracy: Government Forms and Performance in Thirty-Six Countries*, New Haven, CT: Yale University Press.

Lijphart, A. and M. M. L. Crepaz (1991) Corporatism and consensus democracy in eighteen countries: conceptual and empirical linkages, *British Journal of Political Science* 21, 235–56.

Lindblom, C. E. (1977) *Politics and Markets*, New York: Basic Books.

Lipietz, A. (1992) *Towards a New Economic Order: Postfordism, Ecology and Democracy*, trans. M. Slater, Cambridge: Polity Press.

Lowe, P. and J. Goyder (1983) *Environmental Groups in Politics*, London: George Allen and Unwin.

Lowe, P. and S. Ward (1998) Britain and Europe: themes and issues in national environmental policy, pp. 3–30 in: Lowe, P. and S. Ward (eds), *British Environmental Policy and Europe: Politics and Policy in Transition*, London: Routledge.

Lowi, T. (1964) American business, public policy, case-studies, and political theory, *World Politics* 16, 677–715.

Lowndes, V. (2002) Institutionalism, pp. 90–108 in: Marsh, D. and S. Stoker (eds), *Theory and Methods in Political Science*, Basingstoke, Palgrave Macmillan.

Lukes, S. (1974) *Power: A Radical View*, London: Macmillan.

Madison, D. and D. Pearce (1994) *The United Kingdom and Global Warming Policy*, CSERGE Working Paper, GEC 94–27, Norwich:

Centre for Social and Economic Research on the Global Environment, University of East Anglia.

Madison, D. and D. Pearce (1995) The UK and global warming policy, pp. 123–44 in: Gray, T. S. (ed.), *UK Environmental Policy in the 1990s*, London: Macmillan.

Majone, G. (1989) *Evidence, Argument, and Persuasion in the Policy Process*, New Haven, CT: Yale University Press.

Major Energy Users' Council (2001) *Energy Briefing*, April, London: Major Energy Users' Council.

Manners, G. (1995) Energy conservation policy, in Gray, T. S. (ed.), *UK Environmental Policy in the 1990s*, London: Macmillan.

March, J. and J. Olsen (1989) *Rediscovering Institutions*, New York: Free Press.

Marquand, D. (1988) *The Unprincipled Society: New Demands and Old Politics*, London: Jonathan Cape.

Marsh, D. and R. A. W. Rhodes (eds) (1992) *Policy Networks in British Government*, Oxford: Oxford University Press.

McCargo, D. (2004) *Contemporary Japan* (2nd edn), New York: Palgrave Macmillan.

McCormick, J. (1991) *British Politics and the Environment*, London: Earthscan.

McEachern, D. (1993) Environmental policy in Australia 1981–91: a form of corporatism?, *Australian Journal of Public Administration* 52, 173–86.

McGowan, F. (1993) Energy policy in the UK to 1992, pp. 1–14 in: Thomas, S. (ed.), *Energy Policy: An Agenda for the 1990s*, Brighton: Science Policy Research Unit, University of Sussex.

Michaelowa, A. (2003) Germany – a pioneer on earth feet?, *Climate Policy* 3, 31–43.

Midttun, A. and O. Hagen (1997) Environmental policy as democratic proclamation and corporatist implementation: a comparative study of environmental taxation in the electricity sector in the Nordic countries as of 1994, *Scandinavian Political Studies* 20, 285–310.

Ministry of Economy, Trade and Industry and Ministry of the Environment (2002) Enerugî seisaku no minaoshi to dôseisaku niokeru kankyô hairyo no bapponteki kyôka nitsuite [Review of energy policy and drastic strengthening of environmental consideration of the policy], 15 November, www.env.go.jp/policy/info/energy/04.pdf (last accessed 31 July 2005).

Ministry of Land, Infrastructure and Transport (various years) *Rikuun Tokei Nenpô* [Annual report on road transport statistics].

Ministry of the Environment (2004a) 2003 nendo (Heisei 15 nendo) no onshitsu-kôka gasu haisyutsuryô sokuhôchi (Kankyoshô Santeichi nituite), [The 2003 greenhouse gases emissions (provisional calculated by the Ministry of Environment)] November, www.env.go.jp (last accessed 14 June 2005).

Ministry of the Environment (2004b) Kizon enerugî kankei shozei tono kankei ni tsuite, Chûôkankyô shingikai sôgôseisaku/chikyûkankyô gôdôbukai, shisakusôgôkikaku shôiinkai (dai15kai) gijishidai/shiryô [Minutes and materials for the Subcommittee on general policy, Joint committees of general policy and global environment, Central Environment Council (15th meeting)], 10 November, www.env.go.jp/council/16pol-ear/y162-15/mat02_7.pdf (last accessed 14 June 2005).

Mitchell, C. (1998) *Renewable Energy in the UK: Policies for the Future*, London: CPRE.

Mizutani, Y. (1996) Kuruma syakai o toinaosu [Re-questioning the car society], pp. 155–78 in: Takatsuki, H., K. Nakaue and K. Sasaki (eds), *Gendai Kankyôron* [Environmental studies today], Tokyo: Yûhikaku.

Muramatsu, M. (1994) *Nihon no Gyôsei: Katsudôgata Kanryô no Henbô* [Japanese administration: transformation of active bureaucracy], Tokyo: Cyûkô Shinsyo.

Muramatsu, M. and E. S. Krauss (1987) The conservative policy line and the development of patterned pluralism, pp. 514–54 in: Yamamura, K. and Y. Yasuba (eds), *The Political Economy of Japan, Vol. 1: The Domestic Transformation*, Palo Alto, CA: Stanford University Press.

Nakamura, S. and A. Toyonaga (1991) *Making Environmental Policy in the United States and Japan: The Case Of Global Warming*, USJP Occasional Paper, 91–108, Program on US–Japan Relations, Center for International Affairs, Harvard University.

Neary, I. (2002) *The State and Politics in Japan*, Cambridge: Polity Press.

NEF (2001a) Kaigai syuyôkoku shinenerugî no dôkô [Trends in new energy in major countries in the world], 25 December, www.nef.or.jp/info/pdf/development.pdf (last accessed 16 July 2005).

NEF (2001b) Present state of new energy introduction in Japan and its outlook, www.nef.or.jp/english/new/present.html (last accessed 14 June 2005).

New Energy Committee, Advisory Committee for Energy (2000) Sôgô enerugî chôsakai shin enerugî bukai dai11kai gijiyôshi [New Energy Committee, Advisory Committee for Energy, Minutes for 11th meeting, 21 December], www.meti.go.jp/kohosys/committee/oldsummary/0001101/index.html (last accessed 1 August 2005).

New Labour (1997) *Because Britain Deserves Better*, London: Labour Party.

Newell, P. and M. Paterson (1996) From Geneva to Kyoto: The Second Conference of the Parties to the UN Framework Convention on Climate Change, *Environmental Politics* 5, 729–35.

Nippon Keidanren (2002) Kishakaiken ni okeru Okuda kaichô hatsugen no pointo [Points of comments by President Okuda at

press conference], 10 June, www.keidanren.or.jp/japanese/speech/
comment/2002/spe0610.html (last accessed 16 July 2005).

Nishioka, S and T. Morita (1992) Chikyû kikô anteika no tameno
shakaikeizai shisutemu no kôzôtenkan seisaku no taikei – kosumo
puran II [The policy structure for the structural transformation of
socio-economic systems to stabilise global climate – COSMO Plan II],
Environment Research Quarterly 86, 6–13.

Nyenzi, B., T. Ngara, J. Pretel, K. Andrasko, J. Ellis and J. Budhooram
(1999) *Report on the In-depth Review of the Second National
Communication of Canada*, FCCC/IDR.2/CAN, Bonn: UNFCCC,
http://unfccc.int/resource/docs/idr/can02.htm (last accessed 14 June
2005).

OECD (1977) *Environmental Policies in Japan*, Paris: OECD.

OECD (1993) *Environmental Policies and Industrial Competitiveness*,
Paris: OECD.

OECD (1994) *OECD Environmental Performance Reviews: Japan*, Paris:
OECD.

OECD (1995) *Global Warming: Economic Dimensions and Policy
Responses*, Paris: OECD.

OECD (1999) *National Climate Policies and the Kyoto Protocol*, Paris:
OECD.

OECD (2001) *OECD Historical Statistics*, Paris: OECD.

OECD and IEA (2002) *Energy Policies of IEA Countries: The United
Kingdom. 2002 Review*, Paris: OECD.

Ofgem (2001) *Environmental Action Plan*, www.ofgem.gov.uk/temp/
ofgem/cache/cmsattach/8141_factsheet45envactionplan_jul04.pdf
(last accessed 14 June 2005).

Oh, G., M. Reazuddin, P. Schwengels, D. Justus and L. Assunção (1996)
*Report on the In-depth Review of the National Communication of
Japan*, FCCC/IDR.1/JPN, Bonn: UN FCCC, http://unfccc.int/resource/
docs/idr/jap01.pdf (last accessed 14 June 2005)

Ohta, H. (2000) Kyoto giteisho – Sangyô-kai kara no mikata [The Kyoto
Protocol – view from industry], pp. 262–76 in: Takamura, Y. and Y.
Kameyama (eds), *Kyoto Giteisho no Kokusai Seido* [International
institutions of the Kyoto Protocol], Tokyo: Shinzansha.

Okimoto, D. I. (1989) *Between MITI and the Market: Japanese Industrial
Policy for High Technology*, Palo Alto, CA: Stanford University Press.

Olson, M. (1965) *The Logic of Collective Action*, Cambridge, MA:
Harvard University Press.

Olson, M. (1982) *The Rise and Decline of Nations: Economic Growth,
Stagflation, and Social Rigidities*, New Haven, CT: Yale University
Press.

Opschoor, H. and J. V. D. Straaten (1993) Sustainable development: an
institutional approach, *Ecological Economics* 7, 203–22.

O'Riordan, T. and E. J. Rowbotham (1996) Struggling for credibility. The United Kingdom's response, pp. 228–67 in: O'Riordan, T. and J. Jäger (eds), *Politics of Climate Change. A European Perspective*, London: Routledge.

O'Riordan, T. and J. Jäger (eds) (1996) *Politics of Climate Change. A European Perspective*, London: Routledge.

O'Riordan, T., R. Kemp and M. Purdue (1988) *Sizewell B: An Anatomy of the Inquiry*, London: Macmillan.

Panitch, L. (1980) Recent theorizations of corporatism: reflections on a growth industry, *British Journal of Sociology* 31, 159–87.

Paterson, M. (1992) Global warming, pp. 155–98 in: Thomas, C. (ed.), *The Environment in International Relations*, London: Royal Institute of International Affairs.

Pehle, H. (1997) Germany: domestic obstacles to an international forerunner, pp. 161–209 in: Andersen, M. S. and D. Liefferink (eds), *European Environmental Policy. The Pioneers*, Manchester: Manchester University Press.

Pekkarinen, J., M. Pohjola and B. Rowthorn (1992) *Social Corporatism: A Superior Economic System?*, Oxford: Clarendon Press.

Pempel, T. J. (1992) Japanese democracy and political culture: a comparative perspective, *Political Science and Politics* 25, 5–12.

Pempel, T. J. and M. Muramatsu (1995) The Japanese bureaucracy and economic development: structuring a proactive civil service, pp. 19–76 in: Kim, H. K. M. Muramatsu, T. J. Pempel and K. Yamamura (eds), *The Japanese Civil Service and Economic Development: Catalysts of Change*, Oxford: Oxford University Press.

Pempel, T. J. and K. Tsunekawa (1979) Corporatism without labor? The Japanese anomaly, pp. 231–70 in: Schmitter, P. C. and G. Lehmbruch (eds), *Trends Toward Corporatist Intermediation*, New York: Sage.

People's Research Institute on Energy and Environment (1994) *2010 nen Nihon Enerugî Keikaku* [A possible energy option for the future], Tokyo: Daiyamondosha.

Performance and Innovation Unit (2001) *Renewable Energy in the UK – Building for the Future of the Environment*, www.strategy.gov.uk/downloads/files/renewanalytpap1nov.pdf (last accessed 14 June 2005).

Peters, B. G. (1999) *Institutional Theory in Political Science: The 'New Institutionalism'*, London: Continuum.

PETRAS Project (2002) *PETRAS Background Report for Denmark*, www.soc.surrey.ac.uk/petras/reports/denmark.html (last accessed 14 June 2005).

Petroleum Association of Japan (2004) *Annual Review 2003*, Tokyo: Petroleum Association of Japan.

Petroleum Association of Japan (various years) *Sekiyu Gyôkai no Suii* [Trends in the oil industry], Tokyo: Petroleum Association of Japan.

298

References

Pierson, P. D. and R. K. Weaver (1993) Imposing losses in pension policy, pp. 110–51 in: Weaver, R. K. and B. A. Rockman (eds), *Do Institutions Matter? Government Capabilities in the United States and Abroad*, Washington, DC: Brookings Institution.

Putnam, R. D. (1988) Diplomacy and domestic politics: the logic of two-level games, *International Organization* 42, 427–60.

Rate System Committee, Electricity Utility Industry Council (1995) *Chûkan Hôkoku* [Interim report], Tokyo: Rate System Committee, Electricity Utility Industry Council.

Rawcliffe, P. (1995) Making inroads: transport policy and the British environmental movement, *Environment* 37, 16–20 and 29–36.

RCEP (1994) *Transport and the Environment*, 18th report, Cm 2674, London: HMSO.

RCEP (1997) *Transport and the Environment – Development Since 1994*, 20th report, London: HMSO.

RCEP (2000) *Energy – The Changing Climate*, 22nd report, London: TSO.

Reitan, M. (1998) Ecological modernisation and 'realpolitik': ideas, interests and institutions, *Environmental Politics* 7, 1–26.

Rhodes, R. A. W. (1988) *Beyond Westminster and Whitehall: The Sub-central Governments of Britain*, London: Unwin Hyman.

Richardson, J. (ed.) (1982) *The Concept of Policy Style*, London: George Allen and Unwin.

Richardson, J., G. Gustaffason and G. Jordan (1982) The concept of policy style, pp. 1–16 in: Richardson, J. (ed.), *The Concept of Policy Style*, London: George Allen and Unwin.

Road Bureau, Ministry of Construction (2003) *Dôro Poketto Bukku* [Road pocket book], Tokyo: National Road Users' Conference.

Roberts, J., D. Elliott and T. Houghton (1991) *Privatising Electricity: The Politics of Power*, London: Belhaven.

Robinson, C. (1993) *Energy Policy: Errors, Illusions and Market Realities*, Occasional Paper 90, London: IEA.

Robinson, M. (1992) *The Greening of British Party Politics*, Manchester: Manchester University Press.

Rose, R. (1991) Comparing forms of comparative analysis, *Political Studies* 39, 446–62.

Rowlands, I. H. (1995) *The Politics of Global Atmospheric Change*, Manchester: Manchester University Press.

RSPB (2003) *Annual Report 2002–2003*, www.rspb.org.uk/Images/Annual%20Report%202002-3_tcm5-44720.pdf (last accessed 14 June 2005).

Sabatier, P. and H. Jenkins-Smith (eds) (1993) *Policy Change and Learning: An Advocacy Coalition Approach*, Boulder, CO: Westview Press.

SACTRA (1994) *Trunk Roads and the Generation of Traffic*, London: HMSO.

Sawa, T. (1997) *Chikyûondanka o Fusegu* [To arrest global warming], Tokyo: Iwanami Shoten.

Schattschneider, E. (1960) *Semi-Sovereign People: A Realist's View of Democracy in America*, New York: Holt, Rinehart and Wilson.

Schmitter, P. C. (1982) Reflections on where the theory of neo-corporatism has gone and where the praxis of neo-corporatism may be going?, pp. 259–80 in: Lehmbruch, G. and P. C. Schmitter (eds), *Patterns of Corporatist Policy-Making*, New York: Sage.

Schmitter, P. C. (1989) Corporatism is dead! Long live corporatism!, *Government and Opposition* 24, 54–73.

Schmitter, P. C. and G. Lehmbruch (eds) (1979) *Trends Toward Corporatist Intermediation*, New York: Sage.

Schoppa, L. J. (1993) Two level games and bargaining outcomes: why Gaiatsu succeeds in Japan in some cases but not others, *International Organization* 47, 353–86.

Schreurs, M. A. (1995) Policy laggard or policy leader? Global environmental policy-making under the Liberal Democratic Party, *Journal of Pacific Asia* 2, 3–33.

Schreurs, M. A. (2001) Shifting priorities of environmental risk management in Japan, pp. 191–212 in: Social Learning Group, *Learning to Manage Global Environmental Risks. Vol. 1: A Comparative History of Social Responses to Climate Change, Ozone Depletion, and Acid Rain*, Cambridge, MA: MIT Press.

Schreurs, M. A. (2002) *Environmental Politics in Japan, Germany, and the United States*, Cambridge: Cambridge University Press.

Schreurs, M. A., W. C. Clark, N. M. Dickson and J. Jäger (2001) Issue attention, framing, and actors: an analysis of patterns across arenas, pp. 349–64 in: Social Learning Group, *Learning to Manage Global Environmental Risks. Vol. 1: A Comparative History of Social Responses to Climate Change, Ozone Depletion, and Acid Rain*, Cambridge, MA: MIT Press.

Schwartz, F. J. (1998) *Advice and Consent. The Politics of Consultation in Japan*, Cambridge: Cambridge University Press.

Scruggs, L. A. (1999) Institutions and environmental performance in seventeen western democracies, *British Journal of Political Science* 29, 1–31.

Scruggs, L. A. (2001) Is there really a link between neo-corporatism and environmental performance? Evidence for the 1990s, *British Journal of Political Science* 31, 686–92.

Self, P. (1993) *Government by the Market? The Politics of Public Choice*, London: Macmillan.

Shoda, T. (1998) Outlook for introduction of renewable energy sources in Japan and problems in the system for introducing them, *Energy in Japan* 153, 28–46.

Siaroff, A. (1998) Corporatism in twenty-four industrial democracies: meaning and measurement. Unpublished manuscript.

Skea, J. (1993) The environmental dimension, pp. 89–113 in: Thomas, C. (ed.), *Energy Policy: An Agenda for the 1990s*, Brighton: Science Policy Research Unit, University of Sussex.

Smith, M. J. (1993) *Pressure, Power and Policy: State Autonomy and Policy Network in Britain and United States*, Hemel Hempstead: Harvester-Wheatsheaf.

Smith, M. J. (1995) Pluralism, pp. 209–27 in: Marsh, D. and G. Stoker (eds), *Theory and Methods in Political Science*, London: Macmillan.

Social Learning Group (2001) *Learning to Manage Global Environmental Risks. Vol. 1: A Comparative History of Social Responses to Climate Change, Ozone Depletion, and Acid Rain*, Cambridge, MA: MIT Press.

Solesbury, W. (1976) The environmental agenda: an illustration of how situations may become political issues and issues may demand responses from government: or how they may not, *Public Administration* 54, 1–38.

Special Committee for Global Warming Prevention Tax, Central Environment Council (2002) *Cyûkan Hôkoku: Wagakuni ni okeru Ondanka taisaku zeisei ni tsuite* [Interim report: Global warming prevention tax in Japan], www.env.go.jp/council/toshin/t161-h1402/t161-h1402-1.pdf (last accessed 31 July 2005)

Statistical Research and Training Institute, Ministry of Internal Affairs and Communications (2005) *Statistical Yearbook 2005*, Tokyo: Statistics Bureau, Ministry of Internal Affairs and Communications.

Stockwin, J. A. A., A. Rix, A. George, J. Horne, D. Ito and M. Collick (eds) (1988) *Dynamic and Immobilist Politics in Japan*, London: Macmillan.

Study Group on the Economic System to Arrest Global Warming (1992) *Torimatome* [Report], Tokyo: EA.

Subcommittee on Global Environment, Industrial Structure Council (1997) *Hôkokusho* [Report], 12 March, www.meti.go.jp/press/past/h70421rf.html#2-2 (last accessed 1 October 2005).

Subcommittee on Promoting the Spread of Low-Emission Vehicles, LDP Policy Research Council's Environment Division (1996) *Latest Developments Regarding the Low-Emission Vehicle Issue*, 8 March. Tokyo: Subcommittee on Promoting the Spread of Low-Emission Vehicles.

Suwa, Y. (1997) *Nihon wa Kankyôni Yasashiinoka: Kankyô Bijyon Naki Kokka no Higeki* [Is Japan environment friendly? The tragedy of the nation without environment vision], Tokyo: Shinpyôron.

Svendsen, G. T. (2001) Group mobilization and rent-seeking, pp. 17–43 in: Daugbjerg, C. and G. T. Svendsen, *Green Taxation in Question. Politics and Economic Efficiency in Environmental Regulation*, New York: Palgrave.

Takei, M. (1995) Genshiryoku kaihatsu seisaku [Nuclear development policy], pp. 87–132 in: Matsui, K. (ed.), *Enerugî Sengo 50 nen no Kensyô* [Fifty years of energy policy after the war], Tokyo: Denryokushimpôsya.

Takeuchi, K. (1998) *Chikyûondanka no Seijigaku* [The politics of global warming], Tokyo: Asahi Shimbunsha.

Tanabe, T. (1999) *Chikyûondanka to Kankyô Gaikô* [Global warming and environmental diplomacy], Tokyo: Jijitsûshinsya.

Tax Commission (2000) *Wagakuni Zeisei no Genjô to Kadai: 21seiki nimuketa Kokumin no Sanka to Sentaku* [The current national tax system and challenges in Japan: public participation and choice towards the 21st century], www.mof.go.jp/singikai/zeicho/top.htm (last accessed 21 August 2005).

Tegart, W. J. M., G. W. Sheldon, and D. C. Griffiths (eds) (1990) *Climate Change: The IPCC Impacts Assessment*, Canberra: Australian Government Publishing Service.

Thelen, K. and S. Steinmo (1992) Historical institutionalism in comparative politics, pp. 1–32 in: Steinmo, S., K. Thelen and F. Longstreth (eds), *Structuring Politics: Historical Institutionalism in Comparative Analysis*, New York: Cambridge University Press.

Tindale, S. (1996) *Green Tax Reform: Pollution Payments and Labour Tax Cuts*, London: IPPR.

Transport Economics Research Centre (1992) *Chikyûondanka nadono Kanten karano Un'yubumon niokeru Enerugî Taisaku no Arikata ni kansuru Cyôsahôkokusyo* [Report on energy policy in the transport sector from the perspectives of arresting global warming, etc.], Tokyo: Transport Economics Research Centre.

Transport Policy Council (1997) *Un'yubumon ni okeru Chikyûondanka Mondai eno Taiôsaku nitsuite* [Policy to cope with the global warming problem in the transport sector], Tokyo: Ministry of Transport.

Tsebelis, G. (1995) Decision making in political systems: veto players in presidentialism, parliamentarism, multicameralism and multipartyism, *British Journal of Political Science* 25, 289–325.

Tsuchiya, H. (2001) Ondanka-mondai Kaiketsu no tameno WWF Shinario – 2010 nen, 2020 nen Nimuketeno Shihyo [WWF scenario for solving the global warming problem – an index towards 2010 and 2020], www.wwf.or.jp/lib/climate/wwfscenario.pdf (last accessed 13 July 2005).

UK Round Table on Sustainable Development (1997) *Second Annual Report*, London: DOE.

UNU-TERI Initiative on Climate Change (1998) *Report: UNU-TERI Initiative on Climate Change*, United Nations University, Tata Energy Research Institute, www.eva.ac.at/(en)/publ/pdf/unu.pdf (last accessed 14 June 2005).

Vogel, D. (1986) *National Styles of Regulation: Environmental Policy in Great Britain and the United States*, Ithaca, NY: Cornell University Press.

Vogel, D. (1990) Environmental policy in Europe and Japan, pp. 257–78 in: Vig, N. J. and M. E. Kraft (eds), *Environmental Policy in the 1990s: Toward a New Agenda*, Washington, DC: CQ Press.

Vogel, D. (1993) Representing diffuse interests in environmental policy-making, pp. 237–71 in: Weaver, R. K. and B. A. Rockman (eds), *Do Institutions Matter? Government Capabilities in the United States and Abroad*, Washington, DC: Brookings Institution.

von Weizsäcker, E. U. (1994) *Earth Politics*, London: Zed Books.

Wakiyama, T. (1987) The implementation and effectiveness of MITI's administrative guidance, pp. 216–21 in: Wilks, S. and M. Wright (eds), *Comparative Government–Industry Relations: Western Europe, the United States, and Japan*, Oxford: Clarendon Press.

Wallace, D. (1995) *Environmental Policies and Industrial Innovation: Strategies in Europe, the USA and Japan*, London: Royal Institute of International Affairs and Earthscan Publications.

Ward, H. and D. Samways (1992) Environmental policy, pp. 117–36 in: Marsh, D. and R. A. W. Rhodes (eds), *Implementing Thatcherite Politics: Audits of an Era*, Buckingham: Open University Press.

Weale, A. (1992) *The New Politics of Pollution*, Manchester: Manchester University Press.

Weale, A. (1997) United Kingdom, pp. 89–108 in: Jänicke, M. and H. Weidner (eds), *National Environmental Policies*, Berlin: Springer.

Weaver, R. K. and B. A. Rockman (eds) (1993) *Do Institutions Matter? Government Capabilities in the United States and Abroad*, Washington, DC: Brookings Institution.

Weir, M. (1992) Ideas and the politics of bounded innovation, pp. 188–216 in: Steinmo, S., K. Thelen and F. Longstreth (eds), *Structuring Politics: Historical Institutionalism in Comparative Analysis*, New York: Cambridge University Press.

Whitehead, P. (undated) Sustainability, planning guidance, local economy and scheme justification (Torbay and South Devon Friends of the Earth), www.users.zetnet.co.uk/umbrella/foe/transport/ring/phil.htm (last accessed 14 June 2005).

Whitelegg, J. (1989) Transport policy: off the rails?, in: Mohan, J. (ed.), *The Political Geography of Contemporary Britain*, London: Macmillan.

Wilks, S. and M. Wright (1987) Conclusion. Comparing government–industry relations: states, sectors and networks, pp. 274–313 in:

Wilks, S. and M. Wright (eds), *Comparative Government–Industry Relations: Western Europe, the United States, and Japan*, Oxford: Clarendon Press.

Williamson, P. J. (1989) *Corporatism in Perspective: An Introductory Guide to Corporatist Theory*, Beverly Hills, CA: Sage.

Wilson, G. K. (1982) Why is there no corporatism in the United States?, pp. 219–36 in: Lehmbruch, G. and P. C. Schmitter (eds), *Patterns of Corporatist Policy-Making*, London: Sage.

Wilson, J. Q. (1972) *Political Organisations*, New York: Basic Books.

Wilson, J. Q. (1980) *The Politics of Regulation*, New York: Basic Books.

World Commission on Environment and Development (1987) *Our Common Future*, Oxford: Oxford University Press.

WWF-Japan (2003) *2002 nendo Kessan Hôkokusho* [Statement of accounts 2002], www.wwf.or.jp/aboutwwf/accounts/accounts2002. pdf (last accessed 14 June 2005).

WWF-UK (2003) *WWF-UK Annual Review 2002/2003*, www.wwf.org. uk/filelibrary/pdf/areview0203text.pdf (last accessed 14 June 2005).

Wynne, B. and P. Simmons with C. Waterton, P. Hughes, and S. Shackley (2001) Institutional cultures and the management of global environmental risks in the United Kingdom, pp. 93–114 in: Social Learning Group, *Learning to Manage Global Environmental Risks. Vol. 1: A Comparative History of Social Responses to Climate Change, Ozone Depletion, and Acid Rain*, Cambridge, MA: MIT Press.

Yonemoto, S. (1994) *Chikyû Kankyô Mondai towa Nanika* [What are global environmental problems?], Tokyo: Iwanami Shinsyo.

Young, O. R. (1989) *International Co-operation: Building Regimes for Natural Resources and the Environment*, Ithaca, NY: Cornell University Press.

Zhou, D., S. Adejuwon, M. Tichy, M. Everett, A. Achanta and J. Budhooram (1999) *Report on the In-depth Review of the Second National Communication of Australia*, FCCC/IDR.2/AUS, Bonn: UNFCCC, http://unfccc.int/resource/docs/idr/aus02.pdf (last accessed 14 June 2005).

Periodicals

Asahi Shimbun
Asahi Shimbun Evening
Denki Shimbun
Economist
ENDS Report
Enerugî to Kankyô [Energy and Environment]
Environment Daily
Financial Times
Gifu Shimbun
GEN Newsletter
Hansard
Japan Eco Times
Kiko Network Newsletter
Kumanichi Shimbun
Kyoto Shimbun
Mainichi Shimbun
Nihon Keizai Shimbun
Nikkan Kogyo Shimbun
Nikkei Sangyo Shimbun
Times
Yomiuri Shimbun

Interviews

Japan

Kazuo Aichi, Director-General for the Environment (December 1990–November 1991), 18 January 2000.

Mie Asaoka, Representative of the Kiko Forum, 6 January 2000.

Kenji Fujita, Global Environment Department, Planning and Coordination Bureau, EA, 12 January 2000.

Masanobu Hasegawa, Federation of Electric Power Companies, 12 January 2000.

Saburo Kato, Secretariat of the Director-General for the Environment (1989–90), Global Environment Department, Planning and Coordination Bureau, EA (1990–93), 14 January 2000.

Yoichi Kaya, Professor, Keio Gijyuku University, 12 January 2000.

Yasuko Matsumoto, Greenpeace Japan (1990–98), 14 January 2000.

Takamitsu Sawa, Director, Institute of Economic Research, Kyoto University, 10 January 2000.

Britain

Andrew Bennet, House of Commons Committee on Environment, Transport and Regional Affairs (Chairman, 1997–), 2 July 1998.

Tom Burke, Director of the Green Alliance (1982–91), special adviser to the Secretary of State for the Environment (1991–97), 25 June 1998.

Terry Carrington, Environment Directorate, Department of Trade and Industry, 16 April 1998.

Lord Gregson, House of Lords Select Committee on Science and Technology (1980–99), House of Lords Select Committee on Sustainable Development (1994–96), 24 June 1998.

Baroness Hilton of Eggardon, Opposition spokesperson on the environment (1991–97), House of Lords Select Committee on the European Communities, Subcommittee on Environment (1991–95, Chairperson, 1995–98), 20 March 1998.

Michael Massey, Environment Directorate, Department of Trade and Industry, 16 April 1998.

Viscount Mersey, President of the Combined Heat and Power Association (1989–92), 24 March 1998.

David Pearce, personal adviser to the Secretary of State for the Environment (1989–92), 19 May 1998.

Index

310   *Index*